PLASMA ASTRONOMY
and the BIBLE

text by Ellen J. McHenry
based on the writings and presentations of Barry J. Setterfield

Barry's work is based on research by:
Kristian Birkeland
Irving Langmuir
Hannes Alfvén
Anthony Perrat
Halton Arp
Bernard Haisch
Hal Putoff
Timothy Boyer
Luis de la Pena
Marcel Urban
M.E.J. Gheury de Bray
Gerrit Verschuur
Thomas Van Flandern
Yuri B. Kolesnik
C.J. Masreliez
Jean Kovalevsky
A. Montgomery
Lambert Dolphin
and others

For more detailed explanations of the topics in this book, and for a complete list of references, please consider purchasing Barry's book: Cosmology and the Zero Point Energy available at barrysetterfield.org.

© 2023 Ellen J. McHenry
All rights reserved.
Published and printed in U.S.A.
ejm.basementworkshop@gmail.com
ISBN 979-8-9868637-4-0

Retailers: This book can be ordered wholesale at
IngramContent.com

Cover image credit: NASA, ESA, and D. Jones (University of California – Santa Cruz);
Processing: Gladys Kober (NASA/Catholic University of America)

The universe is made of three things: **matter, energy**, and **information**. Matter and energy are physical; information is not. Plasma science is part of the physical realm of matter and energy. Before we begin discussing plasma, we need to define terms that will be used in the rest of the book.

A REVIEW OF ATOMS AND IONS

Atoms are the building blocks of which **matter** is made. There are 118 different kinds of atoms, and each type is called an **element**. You are familiar with many of these elements, such as oxygen, nitrogen, helium, calcium, aluminum, chlorine, iron, copper, zinc, silver, gold, lead, and mercury. Other elements are much less familiar to us, such as iridium, osmium, tantalum, europium or praseodymium. The chart that shows us all the different type of elements is called the **Periodic Table**.

A single atom is made of three types of particles: **protons, neutrons and electrons**. The positively charged protons and the neutral neutrons clump together to make the central **nucleus** of the atom. "Like" charges repel each other, so the positive protons in the nucleus would scatter away from each other were it not for a very strong force, known simply as "**the strong force**," binding them together.

The negatively charged electrons are much smaller than the protons and neutrons; they are so small, in fact, that they add almost no mass to the atom. Electrons are found far away from the nucleus, orbiting (or jiggling) at such a high speeds that scientists can't actually follow them. The more we know about the speed of an electron, the less we know about its position, and the more we find out about its position, the less we know about its speed. This is called the **Heisenberg uncertainty principle**.

The area in which an electron is most likely to be found is called an **orbital**. The name "orbital" seemed like a fitting name back when an atom was believed to look a bit like a tiny solar system with its nucleus as the sun and its electrons orbiting like planets. As we learned more about atoms, the picture changed. Orbitals are now drawn to look like fuzzy balloons. The fuzziness of the shape reflects the fact that we don't know exactly where the electron is. We know the probability of where it is likely to be, but we don't know its exact location. Individual orbitals have more defined shapes:

s   p   d   f

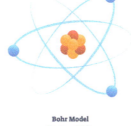

Bohr Model
Electron Orbits

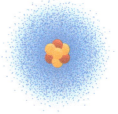

Quantum Mechanical Model
Electron Clouds (Orbitals)

Electron
Negatively charged particles
Atomic mass 0

Neutron
Particles that contain no charge
Atomic mass 1

Proton
Positively charged particles
Atomic mass 1

Electron orbitals are part of a larger structure called a **shell**. We often draw the shells as concentric circles around the nucleus because although this conflicts with the fuzzy quantum picture, it makes the shells easier to understand. The shells are defined by their energy levels with the least energetic being closest to the nucleus. This first shell holds only two electrons. Shells that are farther away from the nucleus hold more electrons and have higher energy levels. Electrons MUST occupy one of these energy levels. Sometimes electrons jump from one shell to another, but they do so seemingly instantaneously. Electrons are never found between the shells. This point will be very important later on.

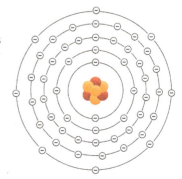

In general, atoms are very strong and stable. This is especially true for smaller atoms such as carbon, oxygen, and nitrogen. Atoms are recycled over and over again as they pass through various cycles such as the water cycle or the carbon cycle. The carbon atoms we use as fuel today (methane, fossil fuels) are the same carbon atoms that ancient plants took in as carbon dioxide.

Even relatively stable atoms can have problems, however. Some atoms have an awkward number of electrons in their outermost shell. A lone ("lonely") electron in an outer shell will actually leave the atom to find a better situation for itself. Before the electron leaves, the atom has an equal number of positive protons and negative electrons. After the electron leaves, the atom's number of electrons will be one less, so the atom will carry an overall positive charge. If the atom gains one or more electrons, it will carry an overall negative charge. Any atom, or a piece of an atom, that carries an electrical charge is called an **ion**. (The word ion comes from a Greek word meaning "to go" because ions were observed going towards electrodes.) Molecules can also become ions if one or more of their atoms gains or loses electrons. The smallest ions are lone electrons and lone protons. A proton can be called a **hydrogen ion**, because a hydrogen atom is made of only one proton and one electron.

The process of becoming an ion is called **ionization**. Most atoms can become **ionized** if enough energy is thrown at them. For example, if you zap a gas with electricity, you can strip electrons away from many of the atoms. Also, high-energy radiation, such as x-rays, can knock electrons off atoms, ionizing them. This isn't good if the atoms being ionized are part of important molecules in your body.

A REVIEW OF ELECTROMAGNETIC ENERGY

One of the most shocking scientific discoveries ever made was when James Clerk Maxwell and Michael Faraday realized that light was only one small part of an entire spectrum of energy waves. Many scientists of that day refused to believe it until enough experiments were done to verify the claim. Today we know that light is, indeed, part of a whole spectrum of electromagnetic energy.

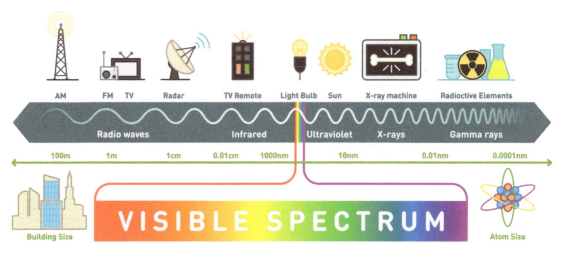

Strange as it may seem, radio waves and light are made of the same "stuff," so to speak. The only difference between them is the length of their waves. As you can see from the illustration, at the low end of the spectrum we have radio waves and microwaves. (Don't confuse radio waves with the sound waves that come out of a radio. Radio waves are the waves that travel from a broadcasting radio station to your radio. Your radio then turns the radio waves into sound waves that you can hear.) At the top of the spectrum we have x-rays and gamma rays. They are called **ionizing radiation** because they contain enough energy to disrupt atoms.

Electromagnetic waves (like all waves) have two essential characteristics: **wavelength and frequency.**

Wavelength, designated by the Greek letter lambda (λ), measures a wave from either peak to peak, or from trough to trough. **Frequency** measures how many times a wave peak goes through a certain reference point in one second. Frequency is measured in units called **Hertz (Hz)**, where one wavelength per second equals 1 Hz.

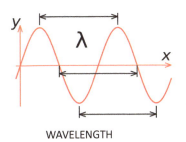

WAVELENGTH

Once an electromagnetic (EM) wave is generated, it will travel on forever unless it is stopped by a substance that can absorb its energy. This ability of the wave to keep going without slowing down is called **self propagation**. Light waves given off by stars can travel all the way across the universe.

In this diagram of an EM wave, the red line represents the electrical portion of the wave and the blue line represents the magnetic portion of the wave. Electricity and magnetism are always found together, and their waves are perpendicular (at a right angle).

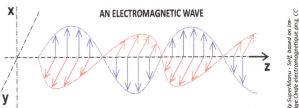

AN ELECTROMAGNETIC WAVE

Sometimes people confuse electromagnetism with electricity. Electricity is made of moving electrons, so it is a type of matter. Electromagnetic waves are not made of matter, but are pure energy. However, moving electrons (electricity) do generate EM waves. Any time that electrons are moving, an EM field is generated. Radio antennas work on this principle. Electricity is sent into an antennae that is specially designed to generate EM waves of a certain wavelength. On the other end, your receiving antenna is specially designed to pick up certain EM wavelengths and turn them back into electrical signals that your radio will turn into sound waves.

All forms of electromagnetic radiation, from radio waves to gamma rays, travel at the same speed. We often call this **"the speed of light"** because light was the first type of EM wave to have its speed measured. The speed of light is represented by the letter "**c**." Scientists have been measuring the speed of light for several hundred years. Their results will be discussed later in the book.

Where does electromagnetic energy come from? In some cases, we see it being generated. Light bulbs make light waves. X-rays are produced by medical machines. Stoves and ovens make infrared heat waves. Stars, including our sun, produce a wide range of EM waves, though we are primarily aware of the light and heat. In all of these cases, the EM waves are being generated by electrons that are jumping between those energy level shells. When an electron falls from a high energy level to a lower one, it releases the extra energy as an EM wave.

The universe is also filled with EM waves whose origin is mysterious. This "background" radiation exists even in a complete vacuum where all matter (including electrons) has been removed. This mysterious electromagnetic energy will play an important role later in the book.

AN INTRODUCTION TO PLASMA

Have you ever played with a plasma ball? It is fascinating to watch the colorful filaments move about, twisting and turning as you place your fingers on various parts of the ball. Sometimes the bright filaments look a bit like lightning, but fortunately they are safely enclosed. What you see happening inside the ball has much in common with galactic phenomena observed by astronomers.

 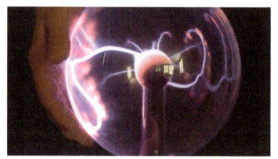

The glass ball is filled with a combination of "inert" (non-reactive) gases: neon, argon, krypton and xenon. These gases belong to a group known as the "noble gases." (On the Periodic Table they are found in the last column on the right. Helium and radon also belong to his group.) The atoms of these gases have outer electron shells that are completely full. Unlike most other atoms, they are not looking to get more electrons or give any way. They are happy the way they are. This makes them a safe target for science demonstrations that involve high-voltage electricity.

The little ball-on-a-stick in the center of the dome is a small-scale Tesla coil that generates high-voltage, low current, high-frequency, alternating-current electricity. Tesla probably hoped his invention would prove to be very useful to industry when he invented it in 1891, but industry moved on and left the Tesla coil to the realm of science experiments and plasma balls. As we learned on the previous page, moving electrons (electricity) will generate electromagnetic waves. In this case, the moving electrons in the Tesla coil generate radio waves having a frequency of about 35,000 Hertz. This is enough energy to seriously disrupt many of the gas atoms. Quite a few of those perfectly happy gas atoms are **ionized** and some of their electrons are scattered throughout the ball. When atoms turn into positive and negative ions, we say that they have turned into **plasma**. Plasma is defined as the fourth state of matter, with the first three being the ones we are familiar with: solid, liquid and gas. Plasma does not have any recognizable shape or form; it's just a soup of atomic ingredients.

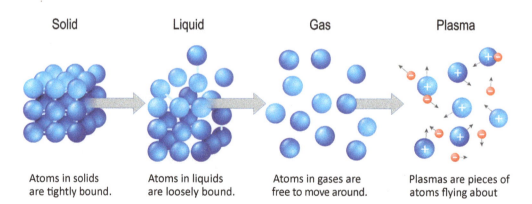

Solid	Liquid	Gas	Plasma
Atoms in solids are tightly bound.	Atoms in liquids are loosely bound.	Atoms in gases are free to move around.	Plasmas are pieces of atoms flying about

As the radio waves stream out towards the edge of the glass dome, they push or pull some of the loose electrons along with them. This is what causes the shape of the filaments. They are streams of charged particles. The color of the filaments comes from electrons that are able to fall back into orbit around a nucleus. As we learned previously, when electrons falls "downward" into a lower orbit, they release their extra energy as EM radiation, often as visible light. The colors of the filaments are unique to these gas atoms. If a scientist looked at the filaments using a tool called a spectrometer, they would be able to identify which gases are in the ball. Each element on the Periodic Table has its own distinct pattern of light, called its **emission spectrum**. The electrons are constantly joining and leaving atoms, so we get a continuous display of color.

Some of the electrons that are streaming outward towards the glass will actually pass through the glass and go off into the air. If we touch the ball, we immediately give the electrons a better option than going into the air. Our electrically conductive bodies offer a "path of least resistance" for the electrons and they will gladly leave the ball through our fingers. We see this concentration of electrons as bright spots under our fingers. This concentration also brightens the filament through which the electrons are traveling.

MIT's Plasma Science Center defines plasma as "super-heated matter, so hot that electrons are ripped away from the atoms, forming an ionized gas." They say it comprises over 99% of the matter in the visible universe, primarily in the form of stars and nebulas. We see plasma here on earth in the form of neon lights, fire, auroras, and lightning.

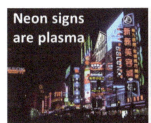
Neon signs are plasma

Flames are plasma

Nebulas Are plasma

Solar Wind is plasma

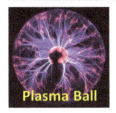
Plasma Ball

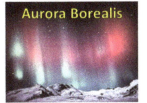
Aurora Borealis

Lightning is plasma

Sun is a plasma

As you can see, plasmas can occur in various forms. Plasmas can be almost **formless,** like the "solar wind" from the sun or a shapeless nebula in outer space. They can also take distinct forms such as **arcs, filaments** and **sheets**. The aurora borealis often looks like sheets of plasma.

Another characteristic of plasma is its mode: it can be in **dark mode, glow mode** or **arc mode**. The plasma that streams out from the sun (the solar wind) is made mostly of single electrons and single protons. As this plasma moves away from the sun, it is in dark mode and can't be seen. When these ionized particles reach earth, they are deflected by the earth's magnetic field (light blue in the diagram) and are drawn in at the poles. As the plasma particles come into contact with the earth's upper atmosphere,

some of the particles collide with atoms of oxygen and nitrogen. The collisions result in much energy being transferred to the gas molecules so that their electrons jump into higher orbits as a result. As the electrons fall back into their orbits, visible light is released. While ionized, the gas molecules act like plasma. Because light is being released, the plasma is in glow mode.

Arc mode is seen in lightning bolts and arc-shaped eruptions on the surface of the sun, and, on a small scale, in the tiny sparks produced by static electricity.

Solar Energetic Particles (Solar Particle Events or Coronal Mass Ejections)

IS PLASMA IN THE BIBLE?

This is not a question about lightning, stars, or fire; obviously those are mentioned in many places in the Bible. What about formless, dark-mode plasma, the kind that exists in outer space, and the kind that secular science tells us the Big Bang produced? Could the universe have started out as nothing but plasma? Is this idea antithetical to the teachings of the Bible?

In Genesis 1:1-2, we read, *"In the beginning, God created the heavens and the earth. And the earth was formless and empty, and darkness was over the face of the deep."* The Hebrew word for "created" is *"bara,"* meaning to create out of nothing (as opposed to creating using pre-existing materials, like creating a chair from wood). The word for "heavens" is *"shamayim,"* and was used for the daytime sky where birds fly, the place where we see stars at night, and the place where God dwells. The word "earth" is *"erets,"* which can have many meanings including country, region, ground, soil, or world. Context is what decides the meaning of the word. Since ancient Hebrews had no words for (not any understanding of) atoms, molecules, matter, plasma, planetary bodies, outer space, or the universe, we can't be surprised that the words "shamayim" and "erets" are used to describe the creation scenario. These were the words they had available.

We are not given any context clues for the use of *"shamayim,"* which has led to much speculation about what it means here. However, we have some very good clues for *"erets."* Nowhere else in the Bible is *"erets"* followed by the description "formless and empty." This is a one-time, unique meaning for the word *"erets."* How formless and empty might the original *"erets"* have been?

The picture on the left shows things that are not formless. Everything from atoms to stars has a distinct form. These cannot be what God made out of nothing in Genesis 1:1. The picture on the right shows plasma. By definition, plasma does not have a form. When it is constrained inside vessels (balls or tubes), or as it interacts with things that do have a form (gas molecules), it can temporarily appear to have a form, but essentially, plasma has no form. The particles of which a completely ionized plasma is made (electrons, protons, quarks) could be used as raw material to make atoms, but they would need an intelligent agent to assemble them. Left on their own, the particles would not turn into the many types of atoms—in the correction proportions and in the right places—required for making living things. The missing ingredient is information. The first sentence of this book stated that the universe is made of matter, energy and information. Plasma gives us matter. We need energy and information to bring the plasma universe to life. Secular science has no reasonable explanation for the origin of information.

The next action in the creation narrative is that "the Spirit of God was hovering over the face of the deep." Here we have a source of intelligence that can bring the plasma particles together to form structured atoms. If God commanded the particles to arrange themselves into atoms, what would have happened next? To form atoms, electrons would need to drop into orbits around newly formed atomic nuclei. What happens when electrons "fall" into orbits? Light is released. Your don't need the sun to have light; all you need is falling electrons.

Can you name another creation story that has light forming before the sun? The Bible's inspired narrative stands out among all other ancient creation stories in this regard. Later in the book we'll also see how the Bible's account of Day 2 matches up very well with what researchers are discovering about the history of the universe.

THE COSMIC MICROWAVE BACKGROUND RADIATION (CMB or CMBR)

Electromagnetic waves can travel on forever unless they run into something that can absorb them. Therefore, those first rays of light that burst forth on Day One could still be crossing the universe. If so, we should be able to detect them. However, they might be in disguise. If anything has happened to those waves since their beginning, such as stretching or compression, they won't look like visible light.

In 1964, Arno Penzias and Robert Wilson began working with a new radio telescope in Holmdel, NJ, that Bell Labs had built to receive radio waves transmitted back to earth by a recently launched NASA communications satellite. This experimental satellite consisted mainly of a huge aluminized plastic balloon. Radio signals aimed at this balloon would be sent from California, and would hopefully bounce off the balloon and then be received by this giant radio telescope. As Penzias and Wilson did preliminary testing of the equipment, they noticed a small amount of static coming in from the microwave range. Unwanted static ("noise") is a common problem for all radio operators; the cause of the static often turns out to be

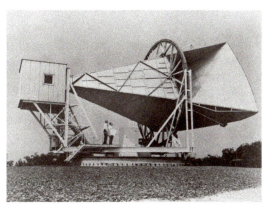

The Holmdel Horn Antenna that Penzias and Wilson used to detect the Cosmic Microwave Background.

something in the immediate environment, such as other electrical devices being operated in the building, ungrounded wires and outlets, or metal pipes. Penzias and Wilson began looking for anything that could be the source of this static. When they found pigeons roosting inside the box-shaped antenna, they felt sure this must be the source of the problem. They spent a day trapping the birds and cleaning pigeon droppings out of the antenna. However, getting rid of the pigeons did nothing to get rid of the static. To further investigate the mysterious source of the microwave static, they pointed the telescope at a patch of empty sky from which they were sure no radio signals would be coming. The telescope still registered the same background level of microwaves. They pointed it at another empty place in the sky. Same results. Every blank place in the night sky gave the same results. They either had a defective telescope or had made a major scientific discovery. Over a decade later, Penzias and Wilson received the Nobel Prize for their discovery of the Comic Microwave Background radiation (CMB or CMBR).

This background radiation had actually been predicted years earlier by cosmologists who were studying the implications of the new Big Bang theory. They reasoned that after being in a very hot plasma state, the original matter would eventually cool enough that particles would clump together to form hydrogen and helium atoms. The creation of these atoms would result in a release of

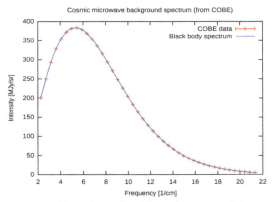

The curved line is the predicted thermal properties of the CMB. The red marks are actual data points. The correlation is stunning.

light (due to electrons falling into orbits). As the universe then expanded, these light waves would be stretched, causing their wavelengths to increase, shifting them down into the range of microwaves or radio waves. In 1948, Ralph Alpher and Robert Herman even estimated that the temperature of this radiation would be about 5° Kelvin. (The Kelvin temperature scale starts at absolute zero which corresponds to -273° C.) Penzias and Wilson measured the CMB to be 4.2° K. Because the actual data was so close to the predicted value, the discovery of the CMB was heralded as "proof" that the Big Bang theory was correct.

The Cosmic Microwave Background has been the subject of much investigation since its discovery. Several satellite missions were designed to take more accurate measurements and make a CMB map of the whole sky. The most recent satellite mission was WMAP from 2001-2010. The map produced by this mission is shown here. The sky is a sphere all around us, not a flat oval, but this CMB map does for the sky what this oval world map does for the globe.

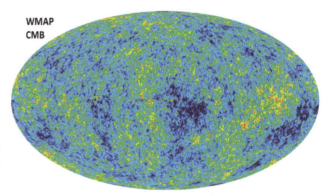

WMAP CMB

This CMB map is color coded so that the warmest areas are in red and yellow, then cooling to green, and finally with blue representing the coldest areas. The difference between the hot and cold areas, however, is extremely small. The most recent estimate of the temperature of the CMB is 2.7255 ±0.0005° K. That means hot areas are 2.7260 and the cold areas are 2.7250, a negligible difference. The temperature of the CMB is remarkably homogeneous.

While the universe was nothing but plasma, energy waves would have bounced around, hitting many of the plasma particles. Imagine trying to hit a cue ball across a billiard table that was filled with over 100 balls. Your cue ball would not go very far. Every time it hit a ball, it would change direction and lose energy. A similar thing happens to light in a bank of fog. The photons of light keep getting scattered by water molecules and are unable to get out of the fog. In the early universe, light was trapped in a "plasma fog." As soon as all the plasma particles turned into atoms, space became transparent and light was able to shine out. This illustration shows an artist's conception of looking back in time, with the pinpoint of light being the Big Bang and the green area representing the formation of the CMB. Some scientists speculate that the universe before the formation of atoms would have been an orange or yellow fog. However, the Bible tells us that "darkness was over the face of the deep."

One of the studies of the CMB was done by a project known as BOOMERanG (Balloon Observations Of Millimetric Extragalactic Radiation and Geophysics). In the year 2000, lead scientists on this project had collected and analyzed enough high-quality data to be able to announce that the universe is nearly flat. This might sound strange, but cosmologists had been debating the nature of the fabric of space and were not sure if the geometry of the universe was curved outward, curved inward, or was flat. BOOMERanG proved that the universe's geometry is flat and thus, very stable.

The BOOMERanG scientists announced another interesting and somewhat surprising result of their research. Professor Paolo di Barnadis of Rome University recognized a pattern in the map of the CMB. It had the mark of something that had been shaped

The BOOMERanG project getting ready to deploy in Antarctica. The telescope flew at an altitude of 42,000 meters.

by sound waves. Prof. di Barnardis said, "The early universe was full of sound waves compressing and rarefying the plasma, much like sound waves compress and rarefy air inside a flute or trumpet. For the first time, the new data clearly show the harmonics of these waves."

The scientists estimated that the speed of the sound waves could have been as high as 57% of the speed of light and their volume could have been about 110 decibels—the same as a loud concert. These incredibly fast and loud sound waves were concentrating the plasma into filamentary structures. The louder the sound, the greater the compression, and the more well-defined the filaments would be. At the conference where these BOOMERanG results were announced, di Barnadis' colleague, Prof. Andrew Lange of CalTech, said, "Using a music analogy, we could tell what notes we were seeing. We see not just one, but three of these peaks, and can tell not only which note is being played but what instrument is playing it. We can begin to hear, in detail, the music of creation."

This graph shows that they found three major peaks in their data. These peaks, oddly enough, correspond to the notes A, F# and C#, which comprise an F# minor chord, although this chord is being sounded at about 50 octaves below middle C. Secular scientists do not know the origin of these sound waves. Some have proposed a system of black holes. For creation scientists, these notes are not hard to explain. God <u>said</u>, "Let there be light." Perhaps these sound waves are the record of God's voice commanding the particles he had just created to begin forming structured matter.

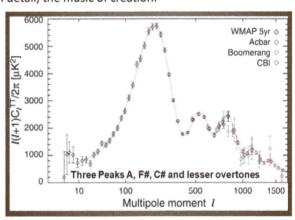

"God spoke, and it was made. He commanded and it stood firm." Psalm 33:9

STAR FORMATION

Before we look at the plasma theory of star formation, let's review the standard "life cycle of stars" saga that we've been taught since grade school.

Orion Nebula

Star formation is said to happen in opaque clumps of very cold gas and dust. The Orion Nebula is often given as an example of a gas and dust cloud that is forming stars. (This nebula is one of the "stars" in Orion's sword.) The gas would be the hydrogen and helium formed at the time of the CMB event. What the "dust" is, or where it comes from, is never well explained. We wonder what elements the dust might be made of. Boron? Carbon? Sodium? Silicon? If so, how were these elements made? (They could not have been made at the same time that the hydrogen and helium were made.) Sometimes the narrative will suggest that the dust could have come from leftover materials wafted over during the formation of a neighboring star, since stars are said to be the "factories" where larger atoms are made. But then where did *that* star get its dust?

Turbulence deep within these clouds is thought to give rise to knots with sufficient mass that the gas and dust can begin to collapse under its own gravitational attraction. As the cloud collapses, the material at the center begins to heat up. A dense, hot core forms and begins gathering more dust and gas. Not all of this material ends up as part of a star—the remaining dust can become planets, asteroids, or comets or may remain as dust. A star the size of our sun requires about 50 million years to mature. Stars can start out as either medium-sized yellow stars, like our sun, or as giant blue stars, like Sirius. Both will eventually turn red. Large stars might go supernova, while small stars likely devolve into white, then black, dwarfs.

Here is an illustration showing another aspect of star formation. Sometimes the gas and dust cloud collapses not into a single star, but into a rotating disc with a very hot spot in the center. The gravitational forces in the hot center collapse it into a star and the remaining debris encircling it eventually condense into orbiting bodies such as planets. This process is sometimes called the Nebular Hypothesis. There are many problems with this hypothesis, but we are going to pass over these for now and keep our focus on the stars.

The "condensing gas cloud" theory of star formation was first studied in depth by Sir James Jeans, a brilliant British physicist who worked in the early part of the 20th century. Enough was known about the behavior of gases at that time for him to come up with equations that could predict the likelihood that a certain gas cloud would be able to condense into a star. The problem with condensation is that as soon as you increase the density of a cloud so that its gravity increases, you also increase its temperature. (Temperature is a measure of molecular motion, and density increases molecular motion.) The increase in temperature will encourage molecules to move away from the dense center, resulting in a decrease in gravity. So it is a "war" between gravity and temperature. How would gravity win?

Recent data about interstellar gas and dust clouds has not solved this riddle. It seems that most gas clouds are unlikely to form stars. The gas cloud needs a "trigger" to make it get dense fast enough that temperature can't win the tug-of-war. Many hypotheses have been put forward as to what might trigger a gas cloud. The most popular mechanism is a supernova. Supernovae explode with enough force that they can create a shock wave that compresses and heats the interstellar gas. If a gas cloud happens to be teetering on the brink of collapse, a sudden shock wave might put it over the brink and cause gravitational collapse. Another idea is that the dust particles (molecules of silicon, ice, CO_2, or even iron) in a dust and gas cloud might be able to absorb and radiate away excess heat created by collision of gas molecules, allowing the temperature to drop just enough for gravity to take over. A third idea is spiral density waves, an idea far too complicated to summarize here.

All of these ideas have a hidden problem: they require stars. Supernovae are exploding stars. Interstellar dust is made of elements that were said to have been made inside stars. Supernovae are suggested as a possible source of the spiral density waves. It's not hard to see that one of the biggest problems for gravitational theories is figuring out how the first stars formed.

The plasma physics theory for star formation does not require pre-existing stars, nor does it depend on gravity. It uses electromagnetic forces that operate the same way at any scale, from tiny wires in a lab up to plasma filaments that are light years across.

By definition, plasma is made of electrically charged particles. When charged particles are in motion, this automatically produces an electric current. Electric currents always have a magnetic field around them, constraining them. The direction of the electric current determines the direction of the magnetic fields. This is often called the "right hand rule," for the reason shown in the diagram.

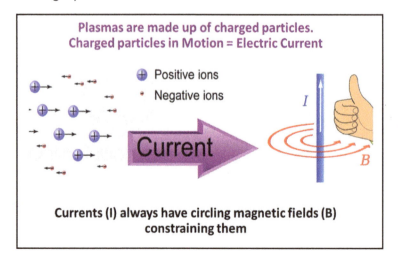

Plasmas are made up of charged particles. Charged particles in Motion = Electric Current

⊕ Positive ions
• Negative ions

Current

I

B

Currents (I) always have circling magnetic fields (B) constraining them

You might wonder how a cloud of plasma in outer space can be moving. However, thanks to the CMB to which we can compare relative motion anywhere in the universe, we now know that everything in the universe in is motion. Also, we don't have to simply assume that interstellar plasma carries an electrical charge—we can actually detect and measure this electricity. Astronomer Gerrit Verschuur announced at the 1999 International Conference on Plasma Science that the interstellar filaments of plasma he had investigated had electrical currents as high as 10,000 billion amperes. If there is electricity, there is also magnetism. The two are linked like the opposite sides of a coin. Magnetic currents coil around electrical currents, constraining them and often shaping them into filaments or sheets.

In the lab, plasma scientists have witnessed what happens when a slight disturbance occurs in a plasma filament. At the point of disturbance, the plasma is suddenly pinched very tightly, and the

pinched spot instantly goes into "arc mode" forming a ball called a plasmoid that glows brightly. They call this the Z-pinch (or Bennett pinch) effect. This pinching happens in about 100 nanoseconds. (A nanosecond is one billionth of a second.)

Does what we see in the lab apply to what we see in outer space? Yes. Everywhere we look in space we see things that look very much like pinched filaments. The ball at the center of these galactic pinches is a very special plasmoid—we call it a star. The star's size and color will be determined by the amount of energy in the filament. Once the star is formed, it will have gravitational properties, but no gravity was required to create it. In M2-9 we also see remnants of another common feature of plasmoids: polar jets.

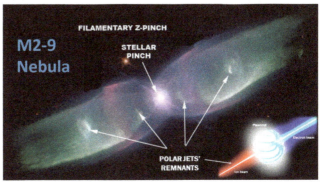

Here we see the Ant Nebula (left) and the Butterfly Nebula (right). Gravitational astronomy struggles to come up with an explanation for their strange shape. If they are not plasma structures, what are they?

There are many other examples of nebulas with symmetric shapes. Not all of them have a star visible at the center, but all of them have shapes that are easily explained by the Z-pinch plasma process. The glowing colors come from hydrogen atoms (red) and oxygen atoms (green) whose electrons have been excited by the high energy of the electric currents flowing through the plasma.

Rotten Egg Nebula

Dumbbell Nebula

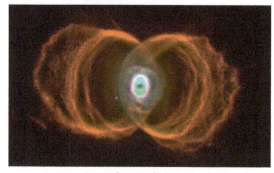

Hourglass Nebula

NGC 2440 Nebula

The star at the center of NGC 2440 is one of the hottest stars ever discovered, at 200,000° C. In all cases, the shapes on either side of the central plasmoid come from the wobble of its polar jets. The greater the wobble, the wider the shape it traces out.

Sometimes many stars can form at the same time; a prime example is located in the constellation Orion. In 2014, the Green Bank radio telescope in West Virginia found a 10-light-year long filamentary structure. An optical telescope image of this structure is shown below on the left. All the tiny white dots on the red filaments are young stars. (The very bright stars are not part of the filament.) Another, similar, filament is show on the right. Plasma astronomers refer to this type of star formation as "beads on a string." If the plasma disappeared we'd wonder how these star ever got lined up.

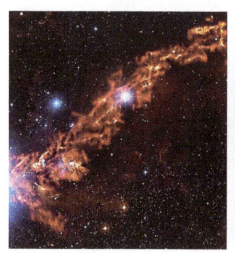

WISE and N(H) map

The star cluster M35 in the constellation Gemini has many stars that appear to be lined up. The filaments are gone.

There are two main types of stars: Population 1 and Population 2. Both can be seen in this photo of the M31 galaxy in Andromeda. Notice that the core of the galaxy has an overall red-orange-yellow appearance, while the spiral arms are predominantly blue. This is because the central region has many cool, red stars and the spiral arms have many hot, blue stars. The central stars are believed to have formed first, and are the older stars. The old stars are classified as Population 2 and the younger as

Population 1. (Yes, it seems backwards.) This classification is not a problem for plasma astronomy. Plasma experiments in labs have shown that plasmoids would indeed form in the center of a spiral shape before they would appear in the arms. (Our sun is Pop. 1 even though it is not blue.)

This classification of stars into two populations does not contradict anything taught in the Bible. Rather, it can demystify some verses, such as Job 38:7. In the translation closest to the original Hebrew, this verse says, "Where were you when I laid the foundation of the earth; when the morning stars were made and all my angels praised me with a loud voice?" Perhaps the morning stars are the older, Population 2 stars, which had already begun to shine before God started forming the earth. The Population 1 stars would be the stars that appeared in the sky on Day 4—the stars that Adam and Eve saw in the sky at the end of the creation week. We will investigate the timing of star formation again, later in the book.

GALAXY FORMATION

Astronomers who are not aware of plasma processes are constantly being surprised by what they find in their telescopes. As telescope technology improves, we are able to see farther and farther into space. Because of their Big Bang assumptions and their reliance on gravitational processes to explain everything, astronomers expect objects at great distances to be small and not very complicated. These objects would have formed very early in the history of the universe, not too long after the Big Bang. Gravitational processes would not have had enough time to form large and/or fully-formed galaxies. Yet this is exactly what they are finding at great distances—large, fully-formed galaxies.

In 2007, the Hubble Space Telescope found a galaxy called A1689-zD1. This galaxy was said to have formed only 700 million years after the Big Bang. Compared to a human life span, this is like photographing someone when they were only a few hours old. The galaxy was so much larger than expected that the discoverers said this find would have a great impact on galaxy formation theories.

In 2014, a discovery was announced by Swinburne University of Technology in Melbourne, Australia. They found mature galaxies with about 100 billion stars at a distance of 12 billion light years. The researchers were stunned and said that the existence of such galaxies "raises new questions."

In 2015, the Hubble telescope found galaxy GN-z11, at a distance of 32 billion light years, said to be equivalent to only 400 million years after the Bang. It is massive and is forming stars quickly—another big surprise for most astronomers.

In 2021, researchers at Cardiff University, Wales, found galaxy ALESS 073.1 at a distance that they thought corresponded to about 10% of the universe's total age. One of the researchers commented that this galaxy looked like an adult when it should look like a child.

A1689-zD1

A1689-zD1 is a fuzzy blur in this square.

Credit: NASA; ESA; L. Bradley (Johns Hopkins University); R. Bouwens (University of California, Santa Cruz); H. Ford (Johns Hopkins University); and G. Illingworth (University of California, Santa Cruz)

Again, the reason that all these finds surprised most astronomers is because according to their gravitational theories, galaxies take billions of years to form. First, clouds of gas and dust collide. The chances of this happening are small, so much time is required. Then, the gas clouds begin to rotate around their common center of mass. They are so large that this also takes a very long time. The rotating motion contracts the gas clouds to form a galactic disk, and then spiral arms form over millions of years. Finally, stars begin to form—another process that is estimated to take millions of years. Gravity can't produce a spiral galaxy within the time frame required by these distant galaxies.

For the plasma physics explanation of how galaxies formed, we will take a look at some lab experiments done by Anthony Perrat, a senior staff member at Los Alamos National Labs, who did his graduate work under Hannes Alfvén, the Nobel Prize-winning Swedish scientist who pioneered the field of plasma physics. Perrat has worked on high-energy-density plasmas, intense particle beams and intense microwave sources, explosively-driven pulsed-power generators, the z-pinch effect, and nuclear fusion target designs.

The colorful diagram shown here on the left is from one of Perrat's experiments with plasma filaments. Each colored strand is an individual filament. As electrical current flows through the filaments, they bunch together as they experience electromagnetic attraction. The illustration on the right shows two filaments from the top view. Red is the electrical current and blue is its magnetic field wrapping around it. In (A), the filaments are far enough apart that their fields stay separated. In (B), the filaments have come close enough together that their magnetic fields have overlapped and combined. This only happens if the electrical currents are flowing in the same direction (note the arrows on the blue circles showing the direction of the magnetic field lines).

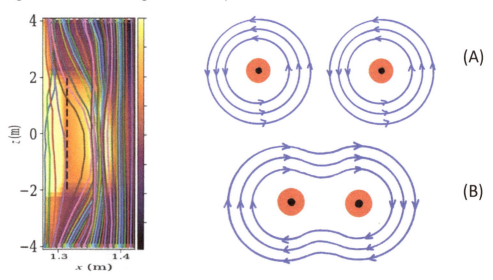

This diagram shows plasma filaments interacting in space. (Electric currents in space are called Birkeland currents, after their discoverer.) If two filaments come very close, their magnetic fields will draw them together. The sketch on the top right shows a cross section of their magnetic field lines. A swirling disc shape will form at the place where the filaments touch.

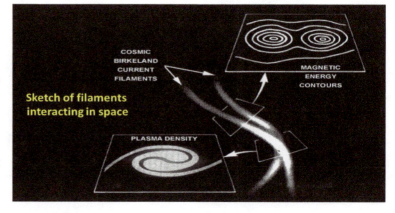

We know that a swirling disc shape will form because Perrat has watched this happen in his lab experiments. This sequence of photographs was taken during one of his experiments.

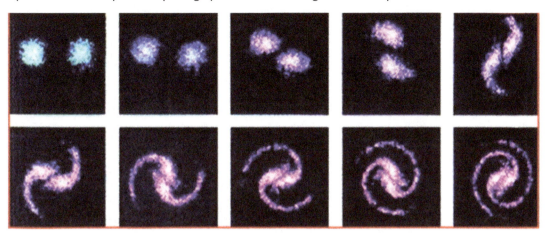

Here are some photos of real galaxies. The similarity is stunning.

| UGC 6093 | NGC 1365 | NGC 1300 |

In his experiments, Perrat discovered that one of the first events to happen in his miniature galaxies was the formation of an ultra-bright plasmoid in the center, which often had "polar jets" coming out of it. (Not visible in the above sequence, but illustrated in the lower left corner of the diagram to the right.) In real galaxies, we call these ultra-bright centers "quasars." They can be so bright that the galaxy around them can't be seen.

The next thing that happened in the lab was that small plasmoids would develop in the central area. These would be equivalent to Population 2 stars. After that, small plasmoids would develop in the spiral arms as the filaments began to fragment. These would be equivalent to Population 1 stars. Thus, the order of galaxy formation suggested by Perrat's experiments is: 1) formation of central quasar and polar jets, 2) formation of stars in central "hub" as spiral arms begin to develop, and 3) formation of stars in the spiral arms.

If Peratt used more than two filaments he found that this would alter the spiral shape somewhat. With enough experimenting, and with the help of a computer simulation of plasma interactions, he found that he could imitate the shape of any real galaxy. He also made a key observation about the spinning behavior of his lab galaxies: the arms rotated at the same speed as the hub, and the strength of the electric current running through the filaments (the Birkeland current) controlled the speed. In the lab, when the current strength dropped off, the spirals slowed down.

Since the 1970s, astronomers have puzzled over the fact that the speed of rotation of the outer arms of most galaxies is the same as the speed of rotation close to their centers. The gravitational theory of galaxy formation predicts that, given the mass distribution we see, the outer edges of a galaxy should be rotating more slowly than the center. In other words, the gravitational explanation only works if the size and number of stars increases as the distance from the hub increases. However, this is not what we see. The outer edges of galaxies do not contain more mass. Thus, astronomers claimed there was "missing mass" in the universe.

In order to preserve the gravitation model of galaxy formation, a rescue device was invented: invisible "dark matter." If galaxies had more mass (than we observe), then the rotation of the arms might be explained by gravity. Thus began the indoctrination of humanity about the existence of dark matter. Dark energy would soon follow. Nothing "dark" is necessary in the plasma physics explanation of galaxy formation. What we observe is in line with what plasma physics predicts. (Even black holes might not be what gravitational astronomy assumes them to be.)

Another surprising and fascinating discovery by Perrat is how fast these plasma processes occur. He kept track of the plasma events in his experiments by using a time code, T, with units equal to 1/10,000 of a second. This diagram lists galactic events with the associated time code of the equivalent lab event. In the lab, a mini-galaxy was complete in 1/4 of a second.

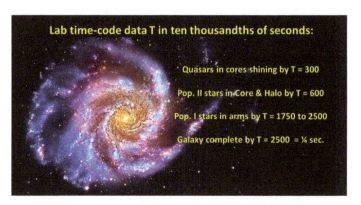

Lab time-code data T in ten thousandths of seconds:

Quasars in cores shining by T = 300

Pop. II stars in Core & Halo by T = 600

Pop. I stars in arms by T = 1750 to 2500

Galaxy complete by T = 2500 = ¼ sec.

Plasma filaments behave the same way, regardless of size, so their behavior in the lab is essentially the same as in outer space. The only difference is the scale factor. Plasma physicists have shown that the scale is a linear one, so a numerical factor can be calculated to "upscale" Peratt's results to galactic proportions. Perrat supplied that scale factor when he published the results of his experiments: 5.87×10^{11} (with the answer being in seconds). In an upcoming section, we will discuss how this scale factor can be used to calculate the actual time it took for galaxies to form, but first let's take a look at the arrangement of galaxies across the universe.

GALAXY DISTRIBUTION ACROSS THE UNIVERSE

In November 2014, the European Southern Observatory (located in Chile) ran an article on their website with this headline: "Spooky Alignment of Quasars Across Billions of Light-years." The astronomers were investigating 93 distant quasars and were shocked to find that quite a few of the quasars' rotation axes were aligned with each other, despite the fact that the quasars are separated by billions of light-years. (A single quasar in shown in the lower left corner of this illustration.) The researchers could not see the rotation axes or the jets of the quasars directly. Instead they measured the polarization of the light from each quasar and, for 19 of them, found a significantly polarized signal. The direction of this polarization, combined with

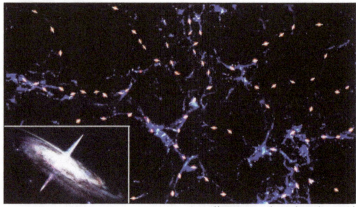

https://www.eso.org/public/news/eso1438/

other information, could be used to deduce the direction of the spin axis of the quasar. The team then extended their research to see if the alignment of the quasars was part of an even larger pattern. To their surprise, they found that quasars are aligned along the massive web of filaments that forms what standard astronomers call the "large scale structure" of the universe.

This diagram is a map of galaxies across the universe. Each dot is a galaxy or a galaxy cluster. Why are galaxies not randomly scattered around the universe? Why this web-like structure? Gravitational astronomers don't know, so they end up using words like "spooky" to describe their discoveries. One of the lead researchers of the ESO team who found the aligned quasars shared this candid thought: "The alignments in the new data, on scales even bigger than current predictions from simulations, may be a hint that there is a missing ingredient in our current models of the cosmos." Yes, indeed—and the missing ingredient is plasma science.

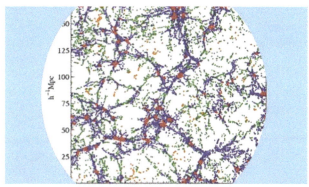

Diagram of the universe by the Tartu Observatory in Estonia

The filamentary structure of the universe is related to the patterns we find in the CMB. Although it is not readily apparent to the untrained eye, computer analysis can confirm that the splotched patterns of the CMB show a positive correlation to the stringy patterns of the large-scale structure diagrams. One event created both. As previously discussed, this event was recorded

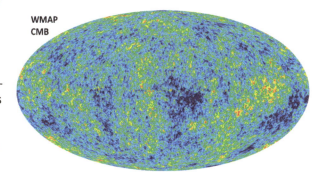

in the Bible as God's voice saying, "Let there be light." At His command, electrons fell into orbits around nuclei, creating stable atoms from the particles in the plasma fog. As God's voice continued to echo throughout the universe, it compressed the remaining plasma into web-like filaments. Many of these filaments would then form quasars, galaxies and stars.

It is said that prediction is the "currency of science." If a theory can make predictions that are then confirmed, our confidence in the theory rises. If observations or experiments prove predictions to be wrong, confidence in the theory goes down.

In 1963, plasma physics pioneer Hannes Alfvén predicted that the universe would be found to have a filamentary structure. In 1991, positions of nearby galaxies were plotted to form this map of our "corner" of the universe. Every dot is the position of a galaxy. (This is just a small portion of the Tartu Observatory's diagram.)

When this diagram was first published, astronomers were shocked. You'd think that they would have been interested in knowing more about a theory that had correctly predicted something so unexpected. But no, most astronomers continued to cling tightly to their gravitational explanations.

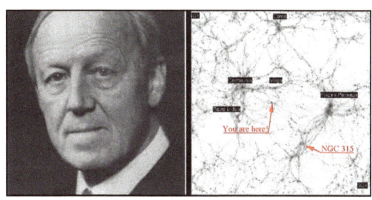

Hannes Alfvén   1991 map of nearby galaxies

DID THE UNIVERSE EXPAND, AND IS IT STILL EXPANDING?

First, why would this thought have crossed anyone's mind in the first place? When we look out at the stars we see a very stable starry landscape. Nothing we observe with our eyes or with ordinary telescopes would make us think that the stars and galaxies are moving farther apart. What caused someone think the universe was expanding?

It began with a discovery made in 1912 by American astronomer Vesto Slipher at Lowell Observatory in Arizona. He had put a spectroscope on his telescope in order to study the light coming from planets. A spectroscope is like a high-tech prism, splitting a light source into its rainbow colors. Each element (hydrogen, oxygen, nitrogen, carbon, etc.) has a unique pattern of colored lines. Often, there are huge gaps between the lines. This illustration shows the patterns (the emission spectra) for hydrogen (top) and helium (bottom). Hydrogen has three bright lines (purple, blue and red) and helium has several lines in each color. No other elements have these patterns. In fact, helium was discovered by pointing a spectrometer at the sun. The scientists saw a pattern they did not recognize and realized they had

discovered a new element. They named it helium, after the Greek word for sun, "helios." It was only later that sources of helium were found on earth. This is a primary reason why astronomers put spectroscopes on their telescopes—to learn what elements are in stars. Stars are made of a number of different elements, so the spectral pattern they make is a blend of all the patterns of the various elements, but astronomers are very good at deciphering the complicated pattern and finding all the individual elements.

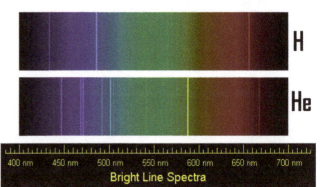

Slipher had been using spectroscopy to study planetary atmospheres, but one night he pointed his telescope at a galaxy. The spectral pattern he saw was definitely recognizable, but he noticed something odd—the pattern was just a little bit off from where it should be. He could see the pattern for helium, for instance, but that bright yellow line hit the nanometer scale a few notches too low. Since the pattern was shifted slightly toward the red end of the spectrum, he said that the pattern was "redshifted." The illustration on the right shows the general idea of red shifting, though actual data would not look this simple.

Slipher wondered if this red shift could be due to something known as the Doppler effect. We've all witnessed the Doppler effect in sounds waves. If a vehicle crosses your path at a reasonably high speed, the sound it makes (especially if has a siren) as it is coming toward you seems to rise in pitch. Then, after it passes and begins to move away from you, the pitch goes down. This is because of what is happening to the sound waves. In picture (A) the car is sitting still and the sound waves going out from it in all directions are evenly spaced. In picture (B), the car is moving to the right. The sound waves in front of it are getting compressed

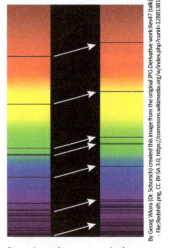

Sometimes the spectrum is shown as black lines on a rainbow background. The arrows show the shift of the lines.

because of the motion of the car. The sound waves behind it are getting stretched out. Since the spacing of the waves controls the pitch, we hear a higher pitch in front of the car and a lower pitch behind the car.

(A)   (B)

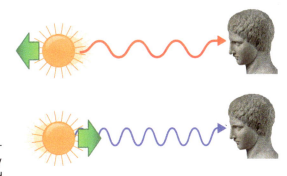

The Doppler effect works with light, also (or any type of EM wave). If a light source is moving away from you, the waves behind it will be slightly stretched. If the light source is moving towards you, the wavelengths will be shorter. Because light waves move so much faster than sound waves, the Doppler effect is not as easy to detect with light. If you watch street lights as you ride along in a car, the lights will look the same regardless of whether you are going toward or away from them. However, with special equipment, and with light sources that are going sufficiently fast at sufficient distances, the Doppler effect can, indeed, be detected. It is a real phenomenon. It is perfectly reasonable for astronomers to hypothesize that galaxy redshift means that galaxies are moving away from us. However, as we will see, the same effect (redshifts) can also have another explanation.

It was quickly accepted as fact that galaxies are moving away from us, but the nature of this movement was still a big question. Were the galaxies actually moving through space, or was space itself expanding, taking the galaxies with it? In 1922, Russian physicist Alexander Friedmann published a set of equations showing that the universe itself might be expanding and estimating the speed at which it might be happening. His work was based on Einstein's general relativity equations. Einstein himself, however, did not think the universe was expanding, but that the galaxies were moving apart in the abundant empty space of the universe.

In 1927, Belgian mathematician George Lemaître published a paper of his own, not directly based on Friedmann's work, but coming to the same conclusion—that the universe itself is expanding. In 1929, Edwin Hubble, of Mount Wilson Observatory in California, did his own research on redshifts. One of his original graphs is shown here. Notice that distance is plotted against velocity, not redshift values. He averaged the results (the straight line) and came up with what is now known as Hubble's Law: *Galaxies are moving away from us at speeds proportional to their distance from us.* In other words, the farther away a galaxy is, the faster it is moving away from us. Hubble estimated that the rate at which this is happening could be as high as 500 km/sec per 3,000,000 light years. In other words, every time the distance doubles, the speed goes up another 500 km/sec.

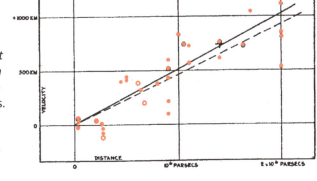

This graph plots velocity versus distance. The actual red shift data is not shown because Hubble has already interpreted it for you.

If you think this rate of expansion sounds extreme, and are not entirely convinced that those red dots are the same thing as a straight line, your intuition does you credit. However, astronomers at that time had less data than we do today, and they did not know of any other way to interpret the redshifts, so they kept walking down this intellectual path.

Telescopes have gotten much better since Hubble's day. The telescope named after him, the Hubble Space Telescope, was launched in 1990 and is still in service. Because the HST is above earth's atmosphere, it has been able to photograph distant galaxies that Edwin Hubble was never able to see. As data came in from very distant galaxies, it became apparent that Hubble's estimate of the expansion rate had been too high. Using Hubble's figures, the distant galaxies that the HST was seeing were moving

away from us faster than the speed of light itself. This breaks a fundamental law of physics (nothing goes faster than light), so something was wrong with Hubble's math. They recalculated and decided that the rate at which galaxies are moving away was more like 60-70 km/sec. per 3,000,000 light years. This patch worked for a while, but the HST kept seeing farther and farther into space, and the redshift data began looking fishy again. Galaxies would be torn apart if they were going as fast as the new data indicated. However, astronomers wanted to keep Hubble's Law, so they strongly endorsed the interpretation that says redshift is due to the expansion of space itself, not to movement of galaxies through space. This interpretation is called "cosmological redshift." If space itself was expanding at a tremendous rate, galaxies might remain intact, though only with the help of hypothetical dark energy.

OTHER WAYS OF MEASURING DISTANCE TO STARS AND GALAXIES

In the 1830s, astronomers used the parallax method to measure the distance to the closest stars, such as Alpha Centauri. This diagram shows how the parallax method cleverly uses the positions of earth as it orbits the sun (A and B) to create the geometry needed for the calculations. This method only works with nearby stars because as the distance increases, the geometric angles get impossibly small to work with.

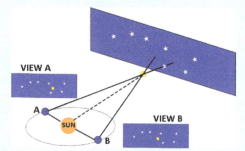

In 1782, British astronomer John Goodricke discovered a star named Delta Cephei (visible above the constellation Cassiopeia) whose luminosity varied over the course of about 5 days. It would dim and brighten with amazing predictability. Its pulsation was timed at exactly 5.366249 days. When more stars like this were found, with pulsations ranging from days to months, a new category of stars was created: "Cepheid variable" stars. In 1908, Henrietta Leavitt was studying photographic plates of Cepheid stars in the Magellanic clouds (dwarf galaxies that orbit the Milky Way). She noticed a correlation between a Cepheid star's pulsation rate (its period) and its luminosity; the longer the period, the greater its maximum brightness. She graphed all her results and developed a mathematical formula to calculate the luminosity simply by observing its period. By comparing parallax distances and apparent luminosities for near Cepheid stars, distance could be approximated for Cepheid stars that were out of range for parallax.

Parallax works for nearby stars less than 300 light years away. The Cepheid variable method picks up right about where parallax trails off and can estimate distances of several million light years. Neither of these methods has anything to do with redshift. They simply compare the relative luminosity of stars. The third method picks up where Cepheid drops off and uses a type of supernova called "1a." These supernovae are extremely bright and produce a very characteristic light pattern. In spiral galaxy NGC 3370 there are several Cepheid variable stars as well as a 1a supernova. Because astronomers had calculated the distance to these Cepheid stars, and they had measured the size of the galaxy, they were able to make an estimate of how far away the supernova was. They recorded how this 1a supernova looked at this distance and could then compare other 1a supernovae to this one. They also looked at how redshifted the light was coming from these supernovae and compared it to the Hubble redshift graph. Thus, the supernovae 1a method became linked to redshift. At extreme distances, only the Hubble redshift method is used.

The idea that the universe is currently expanding has been fed to the public and to astronomy students for decades, so anyone who calls it into question is dismissed as either ignorant or crazy. Hubble's Law has been elevated almost to the status of Newton's Laws. It is rare to find an honest analysis of the situation. Surprisingly, an honest analysis is exactly what we find on a page from one of NASA's websites for kids, called StarChild (noted in bibliography). We read, "*Note that this method of determining distances is based on observation (the redshift in the spectrum) **and** on a theory (Hubble's Law). **If the theory is not correct, the distances determined in this way are all nonsense.**"*) How do we know a theory is not correct? One way is to find data that contradicts it.

In 1976, William Tifft of the University of Arizona published his research on redshifts found in the Coma Cluster (Abell 1656), a group of over 1,000 galaxies located between the constellations Boötes and Leo, and above Virgo. He claimed that the redshift patterns he observed seemed to be "quantized." When something is quantized, it occurs in definite steps or segments. A simple example of quantization is a staircase. The stairs can be labeled 1 to 10. There isn't a step between 1 and 2. It's either step 1 or step 2. You can't be on step 1.5—it doesn't exist.

Coma Cluster

Tifft's claim of quantized redshifts raised many eyebrows in the astronomical community. Why wouldn't there be a steady progression of galaxies getting gradually farther away? If we used a grayscale to represent redshifts, with white for very close to us and black for farthest away, everyone expected to see something like figure (A). What Tifft saw was more like figure (B), with discrete steps.

(A)   (B)

The journal that agreed to publish Tifft's paper felt obliged to include a statement saying that they could not endorse his work, despite the fact that no one could find any mistakes in his techniques or his math.

Astronomers were pretty sure Tifft's results could not be right. The redshifts *must* be gradual, not quantized. However, over a decade went by before anyone attempted to prove that Tifft was wrong. Finally, in 1989, two astrophysicists, Guthrie and Napier, set out to test Tifft's claims. In their own words, they wanted to determine if "this observed effect was due to a diseased radio telescope." Their final results were published in 1997. They concluded that Tifft's work had been "spot on." Guthrie and Napier found the same quantized redshifts in the Virgo Cluster.

This graph is from Guthrie and Napier's 1997 report. The shape of the graph was made by plotting the redshift values of the 48 Virgo spiral galaxies then connecting the dots to make a solid line.

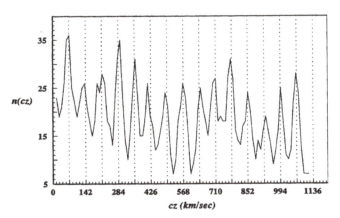

$n(cz)$

The peaks in the graph show where a quantum jump occurs. The dotted lines have been added so you can see the amazing regularity (periodicity) of the quantum jumps. The distance between the peaks is about 71.5 km/sec, which is very close to the value that Tifft had found. Guthrie and Napier were surprised that the same value (about 72 km/sec) was found in two completely different galaxy clusters. However, they said this about their results: "The formal confidence levels associated with these results are extremely high."

Napier, W. M., and B. N. G. Guthrie. *Astrophysics* (1997), 18. pg. 457.

Since then, other studies of redshifts have been done. It was found that a single galaxy could have two redshifts, with the gap between the redshifts cutting right through the center of the galaxy. How could one part of a galaxy be going at a different velocity than the rest? This is impossible! Also, studies of light coming from distant quasars has suggested that the universe may have expanded at first, but is now stable. The quasar light passes through hydrogen clouds on its way to us, and each time it does, the hydrogen absorbs some energy, leaving a black line on the light's spectral pattern. By studying the frequency of these lines, it becomes apparent that expansion is no longer happening.

Quantized redshifts are a big problem for all gravitational explanations of galaxy distribution. But if Hubble's Law is wrong, and the universe is not expanding, then what are the redshifts showing us?

There is another theory that can explain why the redshifts are quantized, but to understand it we'll have to take an excursion into another branch of physics: Stochastic Electrodynamics (SED). The main feature of SED physics is something called the **Zero Point Energy,** or ZPE. Once we understand what the ZPE is and how it would have affected the rate at which plasma processes occurred in the early universe, not only will we have a good explanation for redshifts, but we will also then be able to come to a final conclusion about how long it took for galaxies and stars to form.

EVEN IF THE UNIVERSE IS NOT EXPANDING RIGHT NOW, WAS THERE EVER A TIME WHEN IT DID? AND HOW IS EXPANSION RELATED TO THE ZERO POINT ENERGY?

To make a sweeping generalization, all astronomers and cosmologists, whether secular or creationist, believe that early in its history the universe experienced stretching of some kind. For secular scientists, the expansion idea comes from both the Big Bang theory and the theory of cosmological redshift that we've just discussed. For creationists, the stretching idea largely comes from Bible verses that say that God "stretched out the heavens." (Six verses in the book of Isaiah, two verses in Jeremiah, and one verse in Job, Psalms, and Zechariah.) There has been some disagreement among creationists as to whether these verses suggest a scientific truth or whether they are merely figures of speech, but regardless, these verses have played a significant role. Creationists also point to the CMB as observational evidence of past expansion.

Most young earth creationists believe that the stretching of the universe was one of the events that happened during creation week, probably on Day 1, 2, or 4. Those who favor Day 1 are likely envisioning an event approximately equivalent to the Big Bang's sudden and rapid inflation. The choice of Day 4 makes sense because this was the day that the stars appeared. Day 2 is chosen by those who point out that this is the only day in which an "expanse" was created. Let's explore the Day 2 option. (We'll have to take a few rabbit trails, but we'll come back to the question of expansion.)

God said, "Let there be an expanse between the waters to separate water from water." So God made the expanse and separated the water under the expanse from the water above it. And it was so. God called the expanse "sky." And there was evening, and there was morning—the second day. (Gen. 1: 6-8)

On Day 2, an "expanse" ("*raqia*") was created in the midst of the waters ("*mayim*"). The word "*raqia,*" which is rendered here as "expanse," has also been translated as "firmament" because in other places, "*raqia*" is used to indicate something firm that was spread out, like a blacksmith hammers (spreads out) out metal. However, the *raqia*/firmament is immediately given the name "*shamayim*" which can mean "sky" or "heaven." We again face the problem of limited vocabulary, with one word (*shamayim*) being used for the sky where birds fly, the heavens in which the sun, moon and stars exist, and the place where God dwells. Context must be used to apply the correct meaning.

Historically, the most generally accepted understanding of the expanse of Day 2 has been the sky where birds fly, with the "waters above" being the water vapor in the clouds that falls as rain and snow. This interpretation would have been understandable to the ancients. (Just as a side note, some have speculated that the raqia/firmament correlates to the ancient belief that the sky was a firm dome, and that angelic beings could walk on top of the dome. However, Bible verses that seem to support this idea can have various interpretations; they are always verses that involve poetic metaphors, not literal descriptions.)

a typical illustration of Day 2

There are two problems with this common interpretation. FIrst, there is quite a disparity between what was accomplished on this day compared to the other days. On Day 1, God was shaping the entire universe, and creating quasars, galaxies and population 2 stars—a very big job. It seems strange that on Day 2 he would only put some water vapor into the sky. Secondly, God had not yet created the seas; the earth was still in an ambiguous state without any separtation of land and water. Putting water vapor into the sky (as shown in the Sunday school picture above) seems a bit out of sequence.

In the late 1900s, as creation science blossomed, new ideas were proposed. Some suggested that the expanse between the waters was what we would call "outer space," in which case, there might be a thin layer of water at the very edges of the universe. Or perhaps the "waters above" were in the form of a thin sphere of ice in the upper atmosphere. Another idea was that the raqia/firmament was the crust of the earth, with the "waters below" being located under the crust and "the waters above" including both the sky and the heavens where God lives. If we use plasma astronomy as our starting point, how would this affect our interpretation of Day 2?

There are good reasons to think that the plasma that God created in the beginning was largely composed of hydrogen and oxygen. The most obvious reason is that two hydrogen atoms and one oxygen atom can combine to form a molecule of water, H_2O, and water (*mayim*) is the word the Bible uses to describe the initial matter. Verse 2 says: *"Now the earth was formless and empty, darkness was over the surface of the deep, and the Spirit of God was hovering over the waters."* Some translations say that the Spirit was "driving the waters."

Could the plasma that we observe in the universe today (in filaments and in nebulas) be "leftovers" from the original matter, still in plasma form, that God created out of nothing in the beginning? Are these the building blocks He used to create all atoms? Secular astronomers struggle to come up with a plausible method for elements larger than helium to form. The going theory says that stars are the "factories" where hydrogen and helium nuclei have protons and neutrons added to them to build larger atoms. However, this theory has major technical problems, which astronomers readily admit.

In the 1940s, George Gamov (one of the scientists who correctly predicted the existence of the CMB radiation before it was discovered) spent several years researching and writing about the origin of the chemical elements. He was aware of the problems with adding protons and neutrons to helium nuclei. These "additive" atoms were not stable. Gamov predicted that only one other element would be needed (besides hydrogen) to form atoms of any element. He guessed that this other element might be carbon, but lab testing did not confirm this to be true. Gamov then went on to study and write about other things and decades went by before someone else picked up this line of thought and figured out which element was needed.

In 2010, Edward Boudreaux of the University of New Orleans announced that he had figured out the missing ingredient: oxygen. His calculations showed that if only hydrogen and oxygen nuclei were present in the early universe, all other elements could be formed from these, and in a very short time frame. George Gamov had predicted that once the correct element was found, calculations would show that all other elements could be formed (from hydrogen and this element) in "less time than it takes to cook roast duck and potatoes." Boudreaux's work confirmed this prediction.

Before we go on, we need to note that speculating about all the chemical elements being formed from just two types of ions does not remove the need for a Creator. There had to be just the right amount of each type of atom. Then the atoms had be joined together to form all the molecules necessary for building the earth and encoding information in biological molecules. But perhaps hydrogen and oxygen ions were the raw materials God used in his construction processes.

Astronomers had actually been looking for oxygen ions—in the form of space plasma—for quite some time but had been unable to find them. (To be clear, there was a significant amount of oxygen that they knew about, but it was not nearly the quantity that their theories predicted.) The search for "missing oxygen" was an on-going puzzle. When it was finally discovered, it became obvious why they had missed it for so long. The oxygen ions had only one electron, not the usual 6, 7 or 8. Everyone assumed it was impossible for oxygen to have only one electron, so they were not searching for the "signature" of this type of ion. Once they knew what to look for, they easily found the missing oxygen. It occurred mainly as filaments between galaxies.

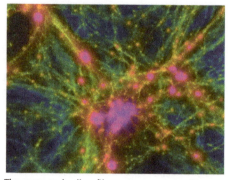
The green and yellow filaments are oxygen plasma.

The discovery of the "missing" oxygen meant that the ratio of hydrogen to oxygen ions in the universe was roughly 2 to 1, the same ratio found in the water molecule. Thus, we might be able to call the original plasma a "water plasma," though complete water molecules may not have been present. Even if there had been other ions in the mix, it still would have qualified as "water plasma."

Surprisingly, many astronomers have not caught up with this recent development in the search for an explanation of how the elements formed. In many texts you will still read about elements being formed by supernovae. However, even if they eventually accept the proposal that hydrogen and oxygen nuclei were used to form the heavier elements, another question remains—where did the original plasma come from? This goes over into the realm of theoretical physics. Quantum physicists are racking their brains as to how a "random fluctuation" in the "space-time continuum" might have generated the first particles of matter. Secular scientists are sure that some day someone will be clever enough to figure out this puzzle and be to explain how something came from nothing. Meanwhile, creationists are not bothered by this question. Genesis 1:1 tells us what we need to know.

As long as matter was in the form of plasma, the nature of the "fabric of the universe" was not critical because plasma is formless and does not have to hold itself together. Atoms are a different story, however. Electrons zoom around the nucleus of their atom, having enough speed to maintain their orbit but not so much speed that they fly out of orbit. How is this balance maintained? Quantum physics asserts various "rules" of how atoms behave, but it doesn't provide an explanation for the rules. There are rules for how electrons can occupy the various shells using the principle of quantization, but the root cause for this is unknown. There's the Heisenberg Uncertainty Principle that says we can't know both an electron's position and its velocity, due to its ability to behave as both a particle and a wave, but why is this so? Physicists began to explore the nature of space itself (3-dimensional space, not outer space). A new branch of physics opened up: Stochastic Electrodynamics (SED).

SED researchers have proposed that atoms are held together by a field of energy waves that are inherent to the vacuum of space. This energy, the Zero Point Energy, is made of waves of electromagnetic

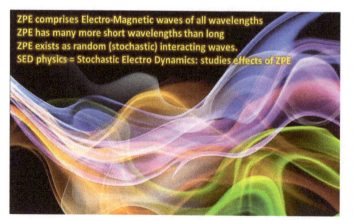

ZPE comprises Electro-Magnetic waves of all wavelengths
ZPE has many more short wavelengths than long
ZPE exists as random (stochastic) interacting waves.
SED physics = Stochastic Electro Dynamics: studies effects of ZPE

energy of all wavelengths, though there are many more shorter waves than longer ones. The waves travel randomly (stochastically) in all directions. If two similar waves happen to run into each other, their combined energy will create a pair of particles, one positive and one negative. As soon as they are formed, the particles immediately slam back together and become energy again. The particles are so short-lived that they barely qualify as existing at all, so they are called "virtual particles."

Virtual particles are possible because of the equation $E=mc^2$. This equation, made famous by Albert Einstein, says that energy, E, is equal to mass times the speed of light squared. In other words, matter can become energy, and energy can become matter. Since the speed of light is currently about 300,000 kilometers per second, E will turn out to be a very large number. (This equation was used to find out how much energy would be released by the small amount of matter in an atomic bomb.) We'll come back to virtual particles, but first we need to learn more about Zero Point Energy waves.

The waves of the Zero Point Energy (ZPE) strike atoms and subatomic particles. SED scientists have estimated that an electron might be struck as many as 18,700 times per orbit around the nucleus. That's an astounding 10^{20} times per second! (Another name for this value is the Compton frequency.) The intense battering of the electron keeps it from flying out of its orbit, as well as keeping it going in its orbit. The ZPE also has an effect on the nucleus, helping to keep it together. **Thus, the ZPE is what maintains the structure of atoms**. It also gives another interpretation of the uncertainty principle: the electrons are being jiggled so much by the ZPE that we can't accurately track their position.

If you imagine ZPE waves as ocean waves, the place where two waves meet, the whitecap, is like a virtual particle—it is there briefly then disappears. These virtual particles have an effect on light waves.

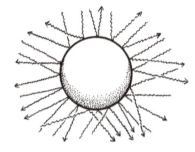

WAVES OF THE ZPE HITTING A PARTICLE

(Note that atomic particles might not be perfect spheres as shown here.)

Bernard Haisch   Hal Puthoff   Timothy Boyer   Luis de la Pena

The ZPE is everywhere, and in everything. SED scientists estimate that there could be as many as 9.3×10^{62} ZPE waves in every cubic inch of space in the universe. This is an immense amount energy. It would be amazing if we could tap into it somehow, but so far, no one has been able to come up with a way to do it. If there is this much energy all around us, why can't we sense it? We don't notice the ZPE for the same reason that we don't notice the 14 pounds per square inch of pressure from the atmosphere. There is as much ZPE/pressure inside us as there is outside us. We live in equilibrium. It is only when the pressure becomes unbalanced that we notice it. (The Eustachian tubes in our ears can make us aware of unequal atmospheric pressure as they "pop" to equalize the pressure.) The ZPE energy is, most significantly, inside our measuring devices, which is why it was so hard to discover. It was actually first discovered by mathematics, not by experiments.

Where do the ZPE waves come from? The electromagnetic waves we interact with are generated as electrons in atoms jump to higher energy levels then fall back down. That is not what is going on with the ZPE. The ZPE energy waves are somehow being generated by the fabric of space itself. One of the current models of the fabric of space looks like a grid work of rotating shapes. The shapes are very hard to explain, as they are not really "made" of anything. The space between them is the smallest space that can possibly exist (estimated to be about $1/10^{35}$ meters). It is impossible to measure anything smaller. If you try to subdivide this space, the space itself begins to unravel and look "choppy." Not all physicists support this model, however. Others prefer

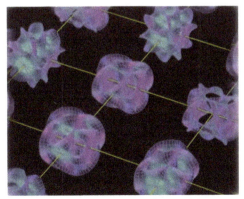

This diagram shows one row of space fabric. The fabric is actually a 3-dimensional grid.

to imagine a vast sea of particle pairs called Planck Particles, named after Max Planck, the founder of quantum physics. It is hypothesized that these particles (which can only be defined as "tiny black holes") come in pairs, with one positively charged and one negatively charged. These particles would rotate around each other. Their electrical charge and their motion would generate the quantum fields in which real particles can exist. If this doesn't make any sense, welcome to the world of theoretical physics. Some theoretical physicists will say that particles of any type don't exist, only fields that have no definite boundaries. Some physicists are coming to the conclusion that when you zoom in to the most fundamental levels, the universe is made isn't matter at all, just fields that store information.

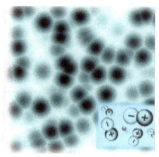

imagined Planck Particle Pairs

No matter which model of empty space you prefer, the idea of stretching can still apply. For the sake of simplicity, let's use the 3-D grid. If the grid work starts out in its smallest and most relaxed form, stretching it can invest it with potential energy, analogous to the energy stored in a stretched rubber band. The potential energy of the rubber band can be turned into kinetic (motion) energy if you release it. SED physicists suggest that if the fabric of space is stretched, this potential energy

Rubber band expanded has energy in its fabric

can be turned into the kinetic energy of the Zero Point Energy (moving electromagnetic waves).

What is the ultimate source of the ZPE energy? What stretched the fabric of the universe? In the case of the rubber band, something, or someone, has to put the energy into the rubber by stretching it. The rubber can't stretch itself. The stretcher has to be "outside of" the rubber band. The same principle holds true for the fabric of the universe. The stretching force must have come from outside of the universe. This is a problem for secular cosmologists, but not for creationists.

On Day 2, God stretched the fabric of the universe, investing it with energy and greatly increasing the Zero Point Energy. The ZPE was necessary for maintaining the structure of the atoms that God would use to make the earth and all living things. (There was a small amount of ZPE on Day 1 to maintain hydrogen atoms, allowing for enough clearing of plasma "fog" so that light could shine out.)

Day 2 began with God setting aside some of the "water" (water plasma?) and designating it as the raw material from which he would make the earth on Day 3. The rest of the "water" (plasma and plasma filaments) was pushed away from the future-earth-plasma-blob, so that the two were separated. A gap (expanse) between them started to grow as God began to stretch the heavens (space). The not-earth stuff (the "waters above") became more and more distant as the expansion continued. It would eventually turn into everything we see in our telescopes. This interpretation of Day 2 turns it from a boring day where hardly anything happened into the day the universe expanded. It doesn't matter if you interpret the "expanse" as the sky where birds fly, or the sky where the sun, moon and stars appear—the expanse eventually turned into both.

Some have noted that Day 2 is the only day on which God did not declare "And it was very good" at the end of the day. We can't be sure of the reason, but perhaps this was because Day 2 was the only day that God did not actually "create" anything; He only expanded what He had created the previous day. Expanding the universe isn't as exciting as making birds and fish, but without it, the atoms in the birds and fish would not have been possible. We've all done big projects that require some preparatory steps before construction actually begins. When land is being surveyed, for example, or when an architect is drawing up plans or when concrete is being prepared by the supplier, it looks like nothing is happening at the construction site; but these preparatory steps are just as important as the actual construction.

At this point we should stop to clarify something that often causes confusion. The stretching that occurred on Day 2 did <u>not</u> stretch the light of Day 1. The fabric of space, whether it turns out to look like Calabi-Yau shapes, little Planck Particle Pairs, or something else, is not "attached" to electromagnetic waves even though the stretching of the fabric creates ZPE waves. Light waves can move through space no matter what is happening to the fabric, in the same way that radio waves can go through the air despite any changes due to wind and weather. Light does not need air molecules in order to travel (unlike sound waves) so whatever is happening to the air molecules does not affect the light waves. Light does not depend on the fabric of space, either, in order to travel, so the stretching of space on Day 2 did not stretch the light.

Another detail that often causes confusion is the relationship between the Day 2 stretching and the creation of the Cosmic Microwave Background radiation. The CMB had technically already formed by the time the Day 2 stretching took place, but the Day 2 stretching dropped its temperature even more. This intense heat at the beginning of Day 1 (at least 6,000° K) came from the collisions between all the particles of matter. As Day 1 progressed, God brought the temperature down just enough to allow neutral atoms like hydrogen for form. To accomplish this cooling, all God had to do was expand the universe a little bit. Expansion always leads to cooling. (This is how your refrigerator works. A pump compresses a gas then allows it to expand rapidly.) As soon as atoms formed, the plasma "fog" was gone and light could shine out. The temperature of the universe was now cooler than at the beginning, but it would drop even more. As God expanded the universe very quickly on Day 2, there was suddenly even more space for that heat to go into. Thus, that original temperature dropped all the way down to what we observe today—just a few degrees above absolute zero.

HOW THE ZPE AFFECTS LIGHT

When two similar ZPE waves collide, their energy combines, and that energy is converted into mass—a pair of virtual particles of some type (up to 21 different kinds have been detected). This is somewhat similar to what happens to ocean waves when they collide. For a brief moment, a white cap is formed. Then it disappears and the waves move on.

Virtual particles are (briefly) made of matter, so, like all matter, they have the ability to absorb a photon of light, effectively stopping the light from traveling any farther. However, a split second later, when the virtual particles slam back together and thus return to being energy, the photon of light is released and can continue on its way. Light going through virtual particles is sort of like a runner going over hurdles. The more hurdles that are in the runner's way, the slower their time will be for the race.

As the ZPE built up over time, more and more virtual particles were being created. Today, the vacuum of space—and the space all around us and inside us—is thick with virtual particles (10^{58} particles per square inch). During creation week and in the first centuries of earth's history, there were far fewer virtual particles, so light waves did not run into as many "hurdles." Thus, light could have traveled much faster in the past. This can help to explain why we can see distant starlight billions of light years away. SED physicists have calculated that without any impedance by virtual particles, light can travel billions of times faster than its current speed.

WHAT EVIDENCE DO WE HAVE THAT THE ZPE IS REAL?

Both quantum electrodynamics (QED, or "quantum physics") and stochastic electrodynamics (SED) use the idea of the Zero Point Energy. For QED, the ZPE is nothing more than a theoretical construct that arises because of the Heisenberg Uncertainly Principle (the observed fact that we can't know both the velocity and position of an electron). QED needs the idea of the ZPE in order to reconcile some of their equations, but they don't believe it actually exists. For SED, the ZPE isn't theoretical—it is made of real waves of real energy. We've already mentioned that the ZPE is in everything, including our measuring devices, which makes it hard to detect. The "Casimir effect" is one of the few experiments that claims to show the presence of ZPE.

Two metal plates (Casimir plates) are put into a vacuum jar, and the air is pumped out until the plates are in an almost perfect vacuum, and the temperature is brought down to as close to absolute zero as possible. The only thing in the jar should be the waves of the

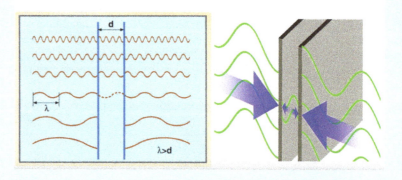

The Casimir Effect

ZPE. The plates are then brought closer and closer together. Electromagnetic waves that have a wavelength greater than the distance between the plates will not be able to exist between the plates. (This is an odd property of electromagnetic waves. If a whole wave (peak to peak) can't get into a space, then the wave won't even attempt to go in.) As the plates get closer, the number of excluded wavelengths increases. As more wavelengths are excluded, the number of waves (and virtual particles) between the plates will decrease. At some point, the number of waves found outside the plates, and the virtual particles they produce, will become enough larger than the number found between the plates that the plates will succumb to the external pressure and suddenly snap together.

QED physics can explain the Casimir effect using their understanding of the way virtual particles work. QED virtual particles are defined as being able to generate waves (de Broglie wavelengths), so, in a way, QED can give a very similar explanation. However, in 2011, C. M. Wilson and his team did a "reverse" Casimir experiment. They brought two plates together very suddenly, which forced out any wavelengths that did not fit. As the plates came together, photons of infrared light came out from between the plates. This result would not be expected in the QED explanation. (For another interpretation of the Casimir effect, see page 43.)

Another experiment that suggests the reality of the ZPE was done at the University of Paris under the direction of Marcel Urban in 2013. He was able to reproduce the current speed of light from the interaction time of photons with virtual particles in a vacuum. The interaction times were independent of photon energies (it worked with any color of light). Dr. Urban wrote that "the quantum vacuum is the origin of the speed of light."

Additionally, the fact that cooling alone will not freeze liquid helium is another strong indication of the reality of the ZPE. Unless pressure is applied, ZPE fluctuations at the atomic level prevent helium's atoms from getting close enough to permit solidification.

Lastly, ZPE fluctuations cause "noise" in electronic circuits, such as microwave receivers, that can never be removed no matter how perfect the technology.

This diagram can help to summarize the lines of thought we have been following. You can see that some ideas overlap with the Big Bang, such something coming from nothing, plasma fog clearing to form the CMB, and a short period of rapid expansion. However, in the model shown here, all of this takes place in a Biblical time frame, not over billions of years.

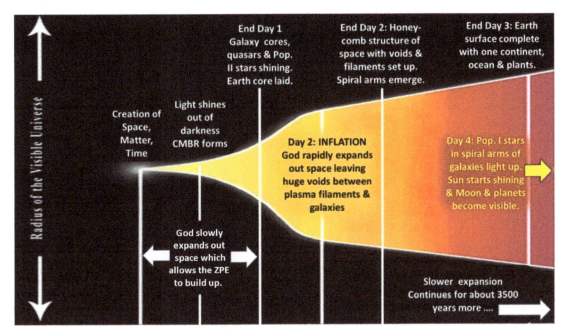

In Big Bang cosmology, there is a short period of inflation (lasting a split second) right after the "Bang," then expansion slows down, then it speeds up again. They have no explanation for why inflation suddenly slowed down. Creationists assume that inflation (stretching) was being carefully controlled by an infinitely wise Creator who knew exactly how much expansion was needed. Notice that after Day 2, the universe continues to expand a bit, but at a much slower rate. Expansion has slowed down so much that today it is barely detectable. It's even possible that it has stopped completely, or even reversed a bit.

Big Bang cosmologists would laugh at the labels in this diagram. To them, it is ridiculous to think that any cosmic process can be accomplished in a matter of days. They tell us with absolute certainty that the universe is billions of years old. Why so old? First, they believe that all cosmic events happen because of gravity. Their ignorance of, or refusal to consider, plasma and electromagnetic processes chains them to gravitational explanations that are extremely slow. When they do calculations based on gravitational processes, their answers come out as millions and billions of years. The other main reason is the redshifting of distant starlight. Few scientists question Hubble's Law or cosmological redshift (stretching of light in an expanding universe). If Hubble's Law is wrong and the redshift is due to something else, and if plasma processes can explain star and galaxy formation, then they really don't know how old the universe is.

At the beginning, before the rapid expansion of the universe, the amount of Zero Point Energy was very low. There was enough ZPE to let hydrogen atoms form and maybe to keep oxygen nuclei together (for water plasma) but that was about it. The whole universe at that time was "low energy." If an energy wave happened to hit an electron in a hydrogen atom, the light re-emitted by the electron as it settled back down would have appeared very red compared to today's measurements of light being emitted by hydrogen atoms. (Today's measurements are called the "laboratory standards" and are the templates against which all astronomical observations are measured. Variance from the standard is interpreted as either redshift or blueshift.) Interestingly, the light from extremely distant quasars looks very red, and is just what we'd expect to see during this low-energy era. This fact will help us to figure out when quasars formed and why we only see them at great distances.

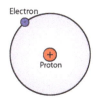

a hydrogen atom

We now need to understand how it is that an increase in the energy of the ZPE causes light to become bluer over time. We have already noted that electrons "jitter" due to the impacts they receive from the ZPE waves and virtual particles. We can measure the intensity of the jitter by using a special device that can hold an electron in a magnetic trap. The trap in this illustration is shown in white. You can't see the electron. Colored light, which is made of many wavelengths, is shone onto the electron, which will scatter light at

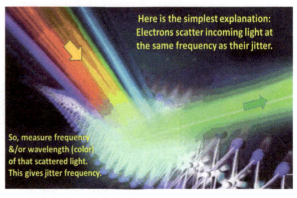

Here is the simplest explanation: Electrons scatter incoming light at the same frequency as their jitter.

So, measure frequency &/or wavelength (color) of that scattered light. This gives jitter frequency.

the same frequency as its jitter. In this illustration, the electron is scattering green light. We can then measure the wavelength (color) of the light and from that we can determine its frequency because wavelength and frequency are related. Thus, we know that electrons are currently being jittered about 10^{20} times per second.

The jitter imparted by the ZPE is a form of kinetic (motion) energy. Since energy can be converted into mass (and vice versa), this increased energy results in an increase in the mass of the electron. Scientists have actually observed the mass of electrons increase over time. (See graph on page 40.) As mass increases, the orbital energy will increase, as well.

To understand why the energy of the orbit would increase, consider what this athlete is feeling as he swings the weight around in a circle. (This is event is called the hammer throw.) In the picture on the left, the athlete is swinging a light weight. In the picture on the right, the athlete is swinging a much

heavier weight. The heavier the weight is, the more energy the athlete will have to expend in order to keep it moving around in a circle. Imagine that the weight at the end of the rope is an electron, and the fixed length of rope is the radius of the electron's orbit. (An atom's "rope" is the electrical attraction between the electron and the protons.) The action of the athlete keeping the weight circling about him is equivalent to the ZPE keeping the electron moving in a stable orbit, not letting it slow down.

As the strength of the ZPE increases, the mass of the electron also increases. For the athlete, this is like the weight becoming much heavier. The athlete must expend more energy to keep the weight moving. Similarly, the orbit of the electron also becomes more energetic. The result of increasing orbital energy will lead to a change in the light that is emitted from the atom.

As discussed previously, these changes in emitted light happened suddenly, producing the quantized patterns we see in the redshift data. The reason for these sudden jumps can be explained by analyzing the wave patterns produced by electrons. Just as light can behave as both a wave and a particle, electrons, which we think of as being only a particle, can also exhibit wave-like properties. We are used to seeing wave patterns in straight lines, but electrons travel in circular orbits, so their wave patterns would be arranged around a circle.

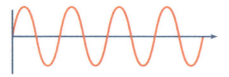
regular wave pattern

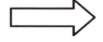

electron wave pattern

It turns out that the wave patterns produced by electrons must fit EXACTLY into their circular orbits. In the examples shown here in red, the wave patterns have varying numbers of "nodes" but they all fit perfectly into the orbit. The example shown in blue, on the right, lets you see what would happen if the wavelengths don't fit perfectly. In this case, the waves simply can't exist.

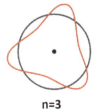    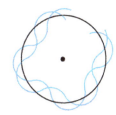

n=3   n=4   n=5   n=6

Thus, there are only a certain number of "permitted" wave patterns. As the ZPE increased, no visible change was seen until the increase in energy reached the critical level where the wave pattern would jump to the next node number. The ZPE would be increasing all the time, but the effect would only be seen at certain points, when it was possible for the orbital to switch to the next pattern. Since wavelength determines the color of light, each time the pattern changed, so would the wavelength of light it emitted—always becoming less red as time went on. Judging by redshift data, little jumps may have occurred over 220,000 times.

Our telescopes act somewhat like a time machine, allowing us to see the state of the universe in times past. On a small scale, we see this phenomenon every day when we look at the sun. Sunlight takes 8 minutes to reach earth, so when we look at the sun we are actually seeing it as it was 8 minutes ago. If the sun exploded right now, we would not know about it until 8 minutes from now. The light from the closest star, Proxima Centauri, is about 4.25 light years away, so when we look at it in the night sky, we are seeing light that left its surface a little over 4 years ago. There are many very distant stars that we can't see with our eyes, even on a dark night in a remote place, but our telescopes let us see them, especially space telescopes like the Hubble and the James Webb. If we can see an object that is 1,000 light years away, we are seeing the object as it looked 1,000 years ago. (This is assuming current light speed.) As we see farther into space we are not only seeing distance, but time, as well. But note that our estimates of cosmic distances are based on the current speed of light. If light was able to travel faster in the past (due to less ZPE), a galaxy that astronomers estimate to be a million light years away might not be showing us what happened a million years ago. We'll draw some conclusions about this on the next page.

Redshift Graph

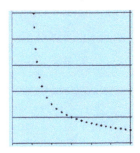

This interpretation of redshifts—a record of how the ZPE has increased since the beginning—gives us a much more plausible explanation for why redshifts increase rapidly as we view ever-more-distant galaxies. The redshifts don't mean that distant galaxies are moving away from us faster than closer ones are. They don't mean that the universe is expanding more rapidly the farther out you go. They don't tell us how far away distant galaxies are. **The redshifts are showing us the strength of the Zero Point Energy over time.**

ZPE Strength Graph

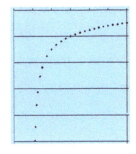

As for the quantization of redshifts, these are *expected* in the ZPE interpretation because they are directly related to something we already know is quantized—the electron energy levels in atoms. Redshifts cutting through galaxies don't pose a problem because this interpretation deals primarily with time, not distance or velocity. If a quasar in the center of a galaxy has a different redshift than the galaxy, this simply means that they came into existence at different times.

HOW LONG DID IT TAKE FOR STARS AND GALAXIES TO FORM?

During the time when the ZPE was increasing, its strength was not even close to what it is today. Without as many ZPE virtual particles in their way, light waves could travel much faster, perhaps as much as 3 billion times faster (according to the calculations of some SED physicists). This means that the light emitted during creation week could easily have crossed the universe in hours or days. The idea that light could have traveled faster in the past doesn't sit well with most astronomers and physicists. The term "light year" would no longer be a valid way of measuring of distance, since it is defined as the distance light can travel in a year. During creation week a light year would have been a very long distance indeed! The idea that the speed of light is not a constant sounds like heresy to most scientists. However, if other constants were also changing while the speed of light was decreasing, all is well. (See the FAQ section for more info.)

We need to allow for a much lower ZPE strength in the early universe. SED calculations show that the ZPE is 3.13×10^9 times higher now that at the time of its origin. Therefore, we must divide by this factor when using Peratt's plasma equations. Peratt calculated that you could multiple his lab results by 5.87×10^{11} and get an estimate of how long plasma processes take on a galactic scale. Peratt used a time code, T, for the steps he observed in his lab. If we use his time codes, then apply both the ZPE correction factor (3.13×10^9) and Peratt's scale factor (5.87×10^{11}), the following information emerges:

The first quasars lit up galaxy cores about 12.5 hours after the origin of the cosmos.
Stars in galaxy cores (population 2 stars) were shining less than 24 hours after the origin.
Stars in the spiral arms of galaxies (like our sun) lit up 3.1 to 4.3 days after the origin.

Our sun is a population 1 star, located in the spiral arms of the Milky Way. The quasar at the center of our galaxy (which is gone now) would have lit up on Day 1, but, according to the time codes Peratt derived from lab experiments, our sun did not light up until Day 4, just as the Bible states.

At Creation, light was 3 billion times faster.

Light from the quasar in our galaxy center reached Earth in 3.7 minutes.

The whole galaxy was lit by the quasar in 15 minutes...
Genesis 1:2-4

Light from the most distant universe got here in less than 10,000 years

Frequently Asked Questions

IF ALL GALAXIES HAVE QUASARS IN THEIR CENTERS (AS SHOWN ON PAGE 32), THEN WHAT HAPPENED TO THE MILKY WAY'S QUASAR?

Our Milky Way galaxy did have a quasar at its center when it was first created, possibly contributing to the light of Day 1. However, as the universe expanded and the ZPE built up very quickly on Day 2, electric and magnetic processes tapered down just as quickly. Most quasars are not emitting brilliant visible light anymore, but some are emitting x-rays and gamma rays. The area that used to be

The light-colored area is the galactic center where Pop.2 stars reside and where the quasar used to be.

the quasar in the Milky Way is now called Sagittarius A* (A-star). Astronomers detect giant "bubbles" of gamma radiation coming perpendicularly out of both sides. They assume this radiation is coming from a black hole because black holes are necessary to gravitational explanations. If the plasma explanation is correct, there is no need to invoke a black hole.

The reason we see bright quasars with huge polar jets in very distant galaxies is because our telescopes allow us to see the galaxies as they were thousands of years ago. If we could see these distant galaxies as they are right now, they would probably look like the Milky Way.

ARE THERE ANY OTHER THEORIES ABOUT WHAT REDSHIFTS MEAN?

Yes. Some plasma astronomers think that redshifts are the result of photons interacting with the vast number of electrons found in large clouds of interstellar plasma. As the light travels through plasma clouds, some of its energy is transfered to the plasma, causing the light to be less energetic. Less energetic light has longer wavelengths, and thus, appears redder. Plasma advocate Halton Arp believed that quasars are ejected from galaxies having intrinsic high redshifts that decrease with time, so redshift is a measure of youth, not distance. Only Barry Setterfield has hypothesized that redshift is directly connected to an increase in the ZPE over time.

IS "THE ELECTRIC UNIVERSE" THE SAME THING AS PLASMA ASTRONOMY?

No. The "Electric Universe" (which is actually trademarked) is based on plasma science and therefore has a lot of overlap with plasma astronomy, but it is not the same. The EU is promoted by a website and YouTube channel called "ThunderboltsProject" (also trademarked). Some of the Thunderbolts videos present aspects of plasma science and are therefore excellent resources for learning about the electromagnetic properties of space. Their videos on the mysterious surface features of the moon and Mars (and how electrical discharge can explain them) are particularly enlightening. However, other videos talk about mythology, not science. They interpret ancient texts and rocks drawings in ways that lead to the conclusion that Saturn was the solar system's original sun, and the planets Mars and Venus were once right next to the earth and loomed large in the sky, and humans witnessed huge electrical discharges between the planets. Some of the EU advocates run a private research project called SAFIRE.

Plasma astronomy/cosmology does not endorse any mythological beliefs. It is simply a branch of science. Some of the pioneer researchers include Kristian Birkeland, Irving Langmuir, Hannes Alfvén, Göran Marklund, and Anthony Peratt. Alfven and Langmuir received Nobel Prizes for their work.

Both EU and plasma astronomy (PA) postulate electric currents in space, but the EU takes the idea several steps beyond the claims of PA. The EU's Electric Sun idea is not exactly the same as PA's claim that the sun is a plasmoid and is therefore electric in nature. Because the Thunderbolts® videos use so many plasma astronomy ideas, it is very easy for the average viewer to confuse EU and PA.

The debunking of the Electric Universe® does not affect plasma astronomy.

IF STARS ARE BALLS OF PLASMA, IS OUR SUN ALSO A PLASMA BALL?

Yes, and this fact can explain a number of interesting observations about the sun that standard astronomers find very puzzling. Here are four "sun puzzles" that mainstream astronomy can't solve:

1) The corona is the outermost layer of the sun, yet it is the hottest layer. The sun's inner atmosphere ranges from 4,000 to 6,000° C, and its outer atmosphere ranges from 10,000 to 1,000,000° C. If the core of the sun is a nuclear furnace where fusion is taking place, shouldn't the hottest places be closer to the core? (One astronomer put it this way: "It is like a flame is coming out of an ice cube.")

2) The center of a sun spot registers a cooler temperature than the surface. If these spots are "holes" that reveal the interior of the sun, and if the interior has nuclear fusion going on, shouldn't the interior of the sun spots reveal the much higher temperatures of the core?

3) Spectrometers detect iron in the sun's corona. If heavy elements like iron are being formed in the sun's core, how do these heavy atoms escape the sun's gravity and end up in the corona?

4) Why does the solar wind speed up the farther out it goes?

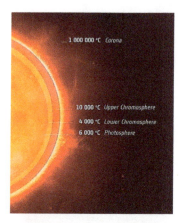

If the sun is a plasma ball, all these strange observations are easily explained. There isn't any nuclear fusion happening in the core of the sun. Astronomers need fusion in stars in order to support their gravitational theories. This assumption is not necessary in plasma astronomy. The sun formed as the result of a pinch in a plasma filament. When plasma filaments are pinched, the resulting plasmoid ball goes into arc mode and begins to glow. This process can happen very quickly and doesn't require millions of years.

The sun spots are showing us exactly what plasma science predicts: the interior of the sun is not hotter than its surface. The corona's extreme heat is created by all the positively charged particles that are being expelled from the surface, creating the "solar wind." As these positively charged particles stream out from the sun's surface, they run into atoms in the sun's atmosphere. The protons are going so fast that the collisions are violent. The violent collisions cause turbulence, which creates intense heat. (You might think of the turbulence that occurs at the bottom of a waterfall as the stream of moving water hits stationary water.)

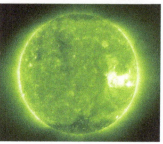

A special filter can show us the element iron (shown as green) in the sun's corona.

Metal atoms, such as iron, that are found in the corona, did not come from the core but were likely created at the surface of the sun. If nuclear fusion is occurring, it is happening at the surface, not in the interior. Plasma scientists have observed the fusion of elements happening on the surface of lab-created plasmoids.

The solar wind picks up speed because its positive charges are being attracted to the negative charges at the outer reaches of the solar system. The farther out you go, the more negatively charged the environment is. This is predicted by plasma science.

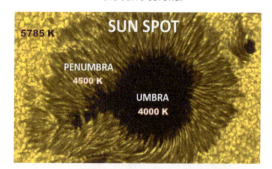

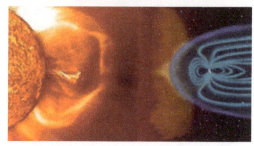

The solar wind blasts plasma particle out into space. The earth is protected by its magnetic field (blue).

IF THE SUN IS A PLASMOID (AND THERE ISN'T ANY NUCLEAR FUSION HAPPENING IN ITS CORE) THEN WHAT IS POWERING IT?

If the sun does not rely on nuclear fusion (hydrogen atoms smashing together to make helium atoms), it needs another source of power to make all that light and heat. Electrical things need a source of electricity. Is the sun "plugged in" to anything? In way, yes it is.

Even plasma astronomers were, at first, not sure where the sun is getting its electricity. Then space probes began sending back information about the density of electrons and other ions in the empty spaces between the planets. It turns out that space actually has a fair amount of mattter in it, even if we can't see it. There can be as many as a billion electrons per cubic meter of space. Now that's still pretty sparse compared the number of electrons packed into a battery, but if you consider how large outer space is, this adds up to a considerable amount. Plasma physicists did calculations to see if this electron density would be enough to keep the sun going, and they found that it is. The sheer volume of electrons makes up for the low density. Thus, the sun has electrons constantly pouring into it from the space around it. Since the solar system is moving through space at about 200 km/sec (149,000 mph) the sun is always moving into new, previously uptapped, sources of electrons. There is plenty of electrical current flowing in outer space to power all the stars, including our sun.

DO BLACK HOLES EXIST?

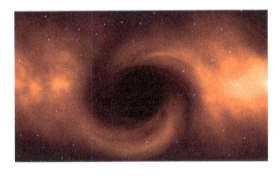

Probably not. The term "black hole" was invented by physicist John Wheeler in 1967 (though based on previous work by Karl Schwartzchild in 1916) in order to explain an unexpected x-ray source in the constellation Cygnus. The energy seemed to be coming out of nowhere, or at least from a very small area. Wheeler remembered Schwarzchild's theory, and then borrowed the mathematical idea of a "singularity" to propose what he called a "black hole."

The idea of a hole that could devour stars and slow down time captured the imaginations of both astronomers and the general public and before long no one questioned their existence. When data came in that didn't support the theory, astronomers merely adjusted "the legend of the black hole" to accommodate the new data. Contradictory data was bound to come along at some point, because black holes are based on the idea that everything can be explained using gravitational processes, when, in fact, it cannot. Recently, astronomers have observed stars forming shockingly close to "black holes." The magnetic fields around "black holes" have been found to be too weak to produce the effects that are seen. A huge blob of gas went right past a "black hole" and, much to the chagrin of astronomers hoping to see a gruesome galactic event, nothing happened. "Black holes" have disappeared then reappeared. Astronomers are willing to ignore these wake-up calls in order to keep their cherished black holes. Wheeler himself did not have very much faith in the idea of black holes. He said this: "The formation of a black hole is equivalent to jumping across the Gulf of Mexico. I would be willing to bet a million dollars that it can't be done, but I can't prove that it couldn't be done."

The x-rays that astronomers are observing are not coming out of black holes. They are coming out of a galaxy's central plasmoid. Electrical currents flow through the filaments found in the spiral arms of a galaxy. The current flows through the arms and into the central plasmoid, and from there, the current goes out in jets, usually on both sides perpendicular to the flat plane defined by the spiral arms. The energy that comes out in the jets depends on the nature of the electrical current that is flowing through the galaxy. If the current is weak, the jets will disappear. When the current increases, the jets will reappear. All of the mysterious behavior of "black holes" is easily explained by plasma science.

HAS A CHANGE IN THE SPEED OF LIGHT BEEN DETECTED IN EXPERIMENTS?

Yes, but barely "in the nick of time." We seem to have caught the tail end of the decline, just before the rate of decline leveled out and has become very hard to detect.

The first person to attempt measuring the speed of light was Galileo. There had been discussion about light speed for several centuries before Galileo but no one had actually set up an experiment to try to measure it. Galileo's experiment seems laughable to us now, but in his day, it seemed reasonable. Galileo's idea was to set up two stations on hilltops that were about a mile away, and use lanterns with shutters that could be opened and closed. One person would quickly open and close the shutter, sending a light signal. The person on the other hill viewing the light signal would wave their hands to indicate that the light had reached him. The hand signal would be visible to those sending the light because they were equipped with a telescope. Needless to say, this experiment was a total flop. Though in a way, it was successful because it caused Galileo and other scientists to realize how incredibly fast light travels.

Some scientists of that day believed that light traveled instantaneously, essentially having an infinite speed. A Danish astronomer named Ole Roemer set out to disprove this idea in the 1670s by using observations of the eclipses of the moons of Jupiter. Data was collected over the course of 12 years, the time it takes for Jupiter to go around the sun once. Roemer found that the times of the eclipses depended on how far away Jupiter was from the earth. The combined motions of Jupiter's orbit and earth's orbit causes the distance between the planets to vary quite a bit. Sure enough, when earth was farthest away from Jupiter, the eclipses occurred later than the times listed on the eclipse charts. When the earth was closer, the eclipses occurred slightly ahead of the predicted times. Roemer correctly perceived that this was due to light traveling longer or shorter distances. Light, he concluded, was very fast, but not infinitely fast. Its speed was measurable. Most scientists accepted this conclusion and the notion of an infinite speed of light soon faded away.

In the 1720s, English astronomer James Bradley tried a new method called aberration. The best way to understand this method is to imagine running through the rain with an umbrella over your head. If you are standing still, the raindrops appear to be falling straight down on your umbrella, but if you are running, the raindrops appear to be coming at you from an angle. The motion of the earth around the sun is like the steady pace at which you are running, and the light from the stars is like the rain. The astronomer measures the angle at which the light is coming down during different seasons of the year, as the earth's position in space changes. Knowing how fast the earth is going and the angle of the light, the speed of light can then be calculated. First, he found that sunlight takes 8 minutes and 12 seconds to reach earth, then he calculated that light speed is about 304,060 km/sec. More recent astronomers have looked at Bradley's records and have concluded that they are amazingly accurate. Then, in 1783, another astronomer used the same method and calculated the speed of light to be about 300,460 km/sec.

During the 1800s, the Pulkovo Observatory in St. Petersburg used the aberration method to measure the speed of light during that century. The same equipment was used throughout the century

and all the scientists were equally adept at using the equipment. Their measurements across the decades of the 1880s went something like this: 300,530 / 300,270 / 299,930 / 299,890 / 299,850 / 299,840 / 299,710 / 299,650 / 299,460 / 299,610 / 299,640 / 299,520 / 299,550 / 277,570. There were also a few measurements that fell slightly above or below these, but several modern statisticians have evaluated these results and have come to the conclusion that there is a genuine downward trend in light speed, even allowing for measurement errors.

Also in the 1800s, another measurement method was being developed using a rotating toothed wheel and some telescopes and mirrors. Armand Fizeau and Alfred Cornu both used this method and came up with measurements of 351,300 km/sec (about 1850) and 300,400 km/sec (1874). A few years later, other scientists examined their work and questioned the functioning of the wheel. They suggested a value of 299,990 km/sec.

The next method used was a rotating wheel with mirrors attached. Variations of this method were used by Leon Foucault, Simon Newcomb, and Albert Michelson. Foucault's estimation was 298,574 km/sec. Newcomb's value was 299,810 km/sec. Michelson improved the original design and in 1883 reported a value of 299,853 (error margin of only 60 km/sec). In the 1920s, Michelson again improved his equipment and was able to come up with very consistent measurements that varied by less than 2 km/sec. His final value was 299,798 km/sec.

In the early 1900s other measuring devices were invented but they all had little issues that affected the precision of their calculations such as moisture in the air, the light going through glass or other media, and in one case even the local ocean tides caused problems.

In 1931, an astronomer named M. E. J. Gheury de Bray made a careful study of all the measurement of light speed from the 1700s through 1929. In 1931, he wrote, "If the velocity of light is constant, how is it that, invariably, new determinations give values which are lower than the last one obtained? There are 22 coincidences in favor of a decrease of the velocity of light, while there is not a single one against it."

In the 1960s, a device called the geodimeter was invented which, among other things, could be used to measure the speed of light. At its heart was the ability to use a crystal to control the emitted light so that its behavior was consistent. The first trials required an instrument at both ends of a line of known distance, but this was soon replaced by a single instrument and a mirror reflector. The graph shown here is a summary of measurements of the speed of light over two decades. Although the very first measurement is anomalously low, the others show a decline.

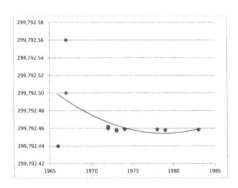

Here is a graph summarizing the measurements of light speed over two centuries. As was mentioned previously, even though measuring devices have become more accurate over the years, when statisticians evaluate data like this, they concur that this is statistically meaningful.

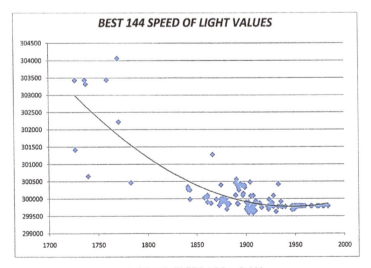

SPEED OF LIGHT ERRORS < 0.1%

In the 1960s, lasers became popular in science labs and it wasn't too long until this method proved to be the most accurate method ever devised. Ken Evenson and his team at the National Bureau of Standards in Boulder developed a frequency synthesis chain which linked microwave output of the cesium frequency standard to the visible part of the spectrum. This made possible the direct measurement of the frequency of a helium-neon laser that was stabilized against a standard transition line in methane (3.39×10^{-6} meters). As a result of their efforts, a new definition of the meter was developed, and was accepted by the 17th General Conference on Weights and Measures in 1983. The new definition read: "The meter is the length of the path traveled by light in a vacuum during a time interval of 1/299,792.458 of a second. **As a consequence of this, the speed of light is now a constant, not to be measured again.**" All physics textbooks now list "c," the speed of light as a never-changing constant equal to 299,792.458.

The variable speed of light had been a topic of much discussion among scientists in the early 1900s. As of 1983, discussion was ended. We now measure the speed of light using a technique that depends on the strength of the ZPE (atomic processes). In essence, we are using a measuring rod to measure a measuring rod. If it is changing (and possibly starting to increase again, suggesting that the universe might be oscillating to maintain stability) we won't know about it.

ARE OTHER "CONSTANTS" CHANGING?

Yes, but only the ones related to the ZPE. The two most important are the rest mass of an electron, and the value of Planck's constant, "h."

The graph shows results of experiments that measured the mass of an electron. Notice that we see exactly the opposite happening here— the value is going up then tapering off. The vertical bars show you the margin of error for each measurement, meaning that even if the value shown is a little off, it should fall somewhere within that bar. (If you are wondering how in the world you weigh something as small as an electron, you don't actually weigh it, of course. You determine an electron's mass by observing how much it is deflected by an electrical field.) Perhaps by now you can figure out why the mass of an electron would be increasing. As the ZPE strength increases, the amount of battering the electrons receive also increases. The electron is slowed down by the extra pummeling it receives from the ZPE. By definition, if an electron slows down, its mass is said to increase.

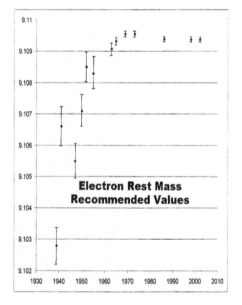

The other value that is directly related to ZPE is "h." This term was first introduced by Max Planck in the year 1900 as way to solve a problem that physicists were having with the analysis of a type of energy called black body radiation. (This problem was jokingly referred to as the "Ultraviolet Catastrophe.") Basically, the theoretical math did not match the experimental data. Planck realized that by adding a constant term to the equation, this "crisis" would be solved. He came at the problem from the theoretical, mathematical side, not realizing that what he had discovered would eventually be shown to be a physical reality. He had discovered the amount of energy in each "stair" on the atomic "staircase." The idea that energy was quantized was new to physics. Planck received the Nobel Prize in 1918.

To measure Planck's constant, scientists have used two different experiments: the "Kibble balance" method, and the X-ray crystal density (XRCD) method. The two methods are different enough that when their results agree, we can be pretty sure of their accuracy. Experimental results (for every constant, not just h), are then submitted to a bureau of standards which then publishes the recommended values for scientists to use. As of 2018, "h" was listed as $6.62607015 \times 10^{-34}$ Joule-seconds. Back in 1985, the accepted value was 6.626176×10^{-34}.

If the Zero Point Energy has increased in strength over time, then the amount of energy in each quanta ("stair") must have changed. This graph shows the results of experiments that measured "h" since 1940. The vertical lines indicated the margins of error. Thus, even if the measured valued should be at the top or bottom of that error bar, we still find that the measured value of "h" has increased over time. It reaches a maximum value in the 1970s, the actually drops a bit. This drop does not invalidate the ZPE interpretation. It merely suggests that the universe has reached maximum size and will now oscillate a bit in order to maintain stability. If the strength of the Zero Point Field begins to diminish, we will see an increase in the speed of light, a drop in the rest mass of an electron, and a drop in the value of "h."

(See bibliography for documentation.)

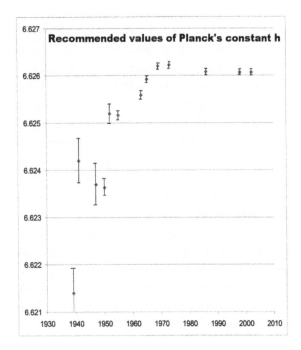

DOES A CHANGING SPEED OF LIGHT VIOLATE THE CONSERVATION OF ENERGY?

No. If "c" changes in the equation $E=mc^2$, energy will be conserved if "m" changes in direct (inverse) proportion to "c^2." We've already looked at graphs that indicate that "c" and "m" have, indeed, changed, with "c" decreasing and "m" increasing. Several notable theoretical physicists have entertained the possibility of a changing light speed, but so far none of them have come up with an improvement on the ZPE explanation. (Some readers might be interested in a book called *Faster Than the Speed of Light*, by João Magueijo, in which he documents his intellectual journey into what he calls VSL (Variable Speed of Light). He and some of his colleagues are seeking a complex mathematical solution.)

The main reason that Einstein chose to pronounce that "c" is constant is because he needed this condition to make his theory of relativity work. Relativity says that there isn't an absolute frame of reference by which we can judge motion; how you see motion depends upon your viewpoint. However, when Einstein formulated his theory, the CMB had not yet been discovered. The CMB does indeed give us an absolute frame of reference for any and all motion in the universe. Thus, the need for light speed to be constant is removed.

For a full discussion that includes equations, this web article might very helpful:
https://tasc-creationscience.org/article/does-changing-speed-light-violate-energy-conservation
Barry Setterfield's book, *Cosmology and the Zero Point Energy*, has an even more in-depth discussion.

IF THE ZPE REALLY WAS LOWER IN THE PAST, DOES THIS HAVE ANY IMPLICATIONS FOR RADIOMETRIC DATING?

Yes. If the ZPE was lower in the past, this would have allowed not only light to go faster but also all atomic processes, including those upon which radiometric dating is based. Radiometric dates are based on the assumption that atoms have always behaved the way they do now. Without this key assumption, we are unable to say how old a rock is. If the SED/ZPE theory is correct, rocks would be able to give radiometric dates of millions or billions of years and still be only thousands of years old.

COULD PLASMA PROCESSES HAVE PLAYED A ROLE IN PLANET FORMATION?

Yes, possibly, but this doesn't mean that the planets were merely the result of natural processes. The earth is fine-tuned for life, which is impossible to explain using only natural forces. However, it is very interesting to note the correlation between what plasma processes predict and what our satellites have detected inside the planets.

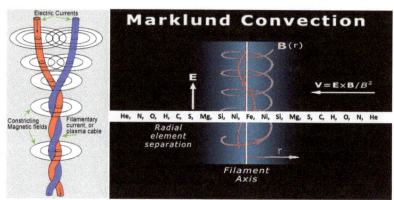

When electric currents are close together, (shown as red and blue lines in diagram on left), a strong magnetic field develops around them, pushing them together to form a filament. Atoms that happen to float past can get caught up into this strong magnetic field and drawn ever closer to the center.

The atoms that are drawn into the filament can be sorted by a process known as Marklund convection. Each type of atom has a unique arrangement of protons, neutrons and electrons. Some arrangements make it difficult for electrons to be stripped away from the atom, and other arrangements make it relatively easy. The relative difficulty of stripping electrons from an atom is called the ionization potential. Atoms with a high potential hold their electrons very tightly. Atoms with a low potential will give them up fairly easily. As the atoms are drawn into the filament, those with the lowest ionization potential, such as iron and nickel, end up in the center, and those with the highest ionization potential (helium, nitrogen, oxygen and hydrogen) end up toward the outside. Atoms such as silicon, carbon and sulfur occupy rings in the middle. This order happens to be approximately what scientists think is inside planets, and possibly even inside stars.

Planets are believed to have cores made of iron an and nickel. Mercury seems to have the largest core compared to its overall volume, and Neptune has the smallest compared to its volume. The size of a planet's silicon-based mantle also varies according to its place in the solar system. Mercury has very little rocky mantle around its large core, Earth and Mars have medium-sized cores and mantles, and the outer planets have tiny cores, very little rocky mantle and are mostly made of gases. This is exactly the proportions that Marklund convection predicts as we go away from the sun.

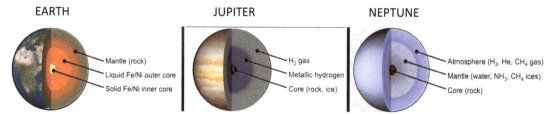

Plasma lab experiments have also revealed that tiny plasmoids (the smaller blue circles in this diagram) can form around the central plasmoid even before the central plasmoid goes into arc mode and lights up. The tiny balls are the equivalent of planets and the central plasmoid is like the star around which they orbit. Thus, planets might be able to form before their stars do (as Genesis suggests).

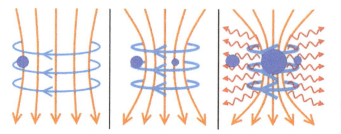

IS SED PHYSICS ACTUALLY A REAL THING?

Yes. The most notable names in SED/ZPE research are: Bernard Haisch, Hal Putoff, Timothy Boyer, Luis de la Pena. They have published their results in peer-reviewed research journals. Also, Marcel Urban, director of the University of Paris, made this statement in the European Physical Journal in January, 2013: "The quantum vacuum is the origin of the speed of light." He and his team were able to reproduce the current speed of light from the interaction time of photons with virtual particles in a vacuum. The reaction times were independent of photon energies. In other words, it did not matter what color of light they used, their results were the same. Urban saw this as proof that the density of the ZPE determines the speed of light.

IS THERE ANOTHER WAY TO INTERPRET THE CASIMIR EFFECT?

Yes. Years after the original Casimir experiments were done, an alternative explanation was found using a force called the van de Waals force. The van der Waals force is a weak attraction between neutral atoms or molecules that don't have any intrinsic electric or magnetic properties. Even in neutral atoms, electrons can be grouped unequally for brief periods of time as they move around the nucleus. If, for a brief (instantaneous) time, the electrons are mostly on one side, the atom will be slightly polar (having one side be more negatively charged). During this brief time, that lopsided atom have a mild attraction to its neighboring atoms. This force is very weak and only exists at very small distances. However, in the Casimir experiment, the plates were brought close enough that the distance was possibly in range of van der Waals forces being able to operate between atoms on opposite plates. Given a sufficent surface area, and nothing holding the plates back, calculations showed that van der Waals forces might be able to pull the plates together. (Calculations using van der Waals forces were about equivalent to those done using ZPE numbers.)

Hrvoje Nicolić wrote a paper (published by Physics Letters in 2016) in which he argued that the vacuum energy (ZPE) explanation does give correct mathematical results, but it does not go far enough into the microscopic details of exactly what is going on. He claims that the van der Waals explanation lies underneath the ZPE explanation.

HAS THE ZERO POINT ENERGY BEEN DETECTED EXPERIMENTALLY?

SED physicists believe so, but QED physicists are still skeptical. The Casimir effect isn't a "slam dunk" for SED. There is an alternate explanation apart from SED. However, in 2013, Finnish researchers were able to "make light appear from nothing" using an experiment where the speed of light was slowed down by making it go through a medium (like light will slow down going through glass or water) while inside a complete vacuum at very close to absolute zero temperature. They detected entangled photons, which aligns well with SED's interpretation of the Casimir effect. Also, in 2021, researchers at Dartmouth used a diamond with embedded light sensors to detect entangled pairs of photons in a cold vacuum—a result they interpreted as "photons coming from nowhere" out of a complete vacuum. This validates the idea that virtual particles are real.

For more information on these experiments, see the bibliography for web addresses.

ARE PLASMA ASTRONOMERS FAMILIAR WITH SED PHYSICS?

As a general rule, not really. Plasma astromy and SED physics are separate branches of science and very few people study both. Barry Setterfield might be the only scientist who has combined them to make a unique cosmology.

IS THERE ANY GENUINE DOPPLER EFFECT HAPPENING TO STARLIGHT?

Yes. When viewing nearby galaxies, we do see Doppler effects. Because the CMB gives us a frame of reference for any motion in the universe, we know that galaxies are moving. Our own Milky Way galaxy appears to be traveling about 300-600 kilometer per second headed in the direction of the Hydra constellation. Some nearby galaxies look redshifted (moving away from us) and others appear blueshifted (moving toward us). When astronomers do their cosmological redshift calculations they have to subtract out any genuine redshifts or blueshifts.

WHY WAS THE LARGE-SCALE (FILAMENTARY) STRUCTURE OF THE UNIVERSE SURPRISING TO ASTRONOMERS?

The Big Bang should have created a universe that was essentially uniform in density. Any variation from homogeneity (uniformity) would have been caused by tiny random fluctuations here and there. We know what random looks like, and the structrure of the universe definitely does not look random. It looks like an organized web where all the parts are connected.

WHY ARE ASTRONOMERS ALWAYS SHOCKED AT THE DATA COMING BACK FROM THE JAMES WEBB SPACE TELESCOPE?

The Big Bang model of cosmology is based on gravity and therefore predicts that billions of years were required for galaxies to form. Looking further into space is like looking further back in time. The JWST is said to be able to look back to as early as only a few hundred millions years after the Big Bang. At this early date, there had not been enough time for galaxies to have formed by gravitational processes. Yet the JWST is returning images of galaxies from this early time period. Headlines proclaim that astronomers will have to completely rethink their theories. But will they? Most of them will not.

DOES PLASMA ASTRONOMY PROVIDE A BETTER EXPLANATION (THAN THE BIG BANG DOES) FOR ALL THE ANGULAR MOMENTUM IN THE UNIVERSE?

Yes. In the Big Bang, we have matter being spewed outwards from a single point. The motion is linear in all directions. Imagine how surprising it would be to examine the aftermath of a dynamite explosion in a rock quarry, and find that all the chunks of rock lying on the ground were spinning around like tops. Perhaps there could be one or two flukes where the rocks happen to hit something at the right angle, but ALL of them? It is just as surprising to look out into the universe and see everything spinning—stars, planets, and galaxies. Everything rotates! Where did all this rotational energy come from? Big Bang cosmology tries to use gravitational collapse (combined with a lot of wishful thinking) to get dust clouds spinning. If they program their computer models just right, they can make it look like it might work, given billions of years.

Plasma science, on the other hand, provides a natural way for spiral motion to start. Birkeland currents (electrical currents in space) naturally twist together, creating spiral motion. Spiral motion has been observed in laboratory experiments with plasma filaments. Also, the most basic rule of electro-magnetism, the "righthand rule," states that an electrical current will automatically generate a curling magnetic field. Of course, we must also allow for the possibility that God simply commanded things to start revolving and rotating. The option we must definitely rule out is the idea that all the rotating and revolving in the universe could have come from the Big Bang.

WHAT ABOUT THE THEORY OF RELATIVITY? WAS EINSTEIN WRONG?

First, we need to consider why the theory of relativity was introduced in the first place. At the beginning of the 20th century, the prevailing view among scientists was that space had to be filled with some kind of medium through which light traveled. They knew that sound waves, and other pressure waves, can't travel through a vacuum; they must travel through solids, liquids or gases. Therefore, they reasoned that electromagnetic waves must also travel through some type of medium, though not solid, liquid or gas. Though they did not know what this medium was, they came up with a temporary name for it: ether. It was assumed that this ether was generally "at rest" throughout the universe, and therefore planets drifting through space would be disturbing the ether, much like a moving car disturbs the air around it. Since the earth travels through space at about 30 km/second, this should be fast enough for an experiment to be able to detect "ether drift" around the earth.

In 1887, Michelson and Morley performed an experiment that was designed to determine if "ether drift" was a real thing. They measured the speed of light in different directions, both with, and against, the direction of the proposed ether drift. What they found was that the detected amount of "drift" was so small that it was within the margin of error due to the limitations of their equipment. Michelson concluded that, "the hypothesis of the stationary ether is thus shown to be incorrect." Some scientists were then ready to give up the idea of ether, but others, including Einstein, did not want to abandon the idea. Thus, an explanation was needed as to why the Michelson-Morley experiment did not detect any ether drift.

A number of proposals were made. The first was by Lorentz in 1904, immediately followed by Einstein's theory of special relativity (SR) in 1905. Einstein explained the M-M results by proposing that 1) there was no absolute frame of reference anywhere in the universe, 2) that the speed of light was an absolute constant, and 3) that mathematical transformations had to be applied to time, space, and mass *both ways* when comparing two objects in motion. By contrast, Lorentz's theory (LR) did not have the restriction regarding an absolute frame of reference, did not need the speed of light to be constant, and proposed that mathematical transformations only apply one way, which meant that clocks, meter sticks and momentum are affected, but time, space and matter are not.

SR requires that time itself be affected by velocity or gravitational potential. LR says that nothing happens to time itself, just to certain types of clocks attempting to keep track of time. SR says that not only is time affected by velocity, but mass and length, as well. LR does not put these requirements on mass or length with respect to velocity.

Einstein used his required length transformation of space to explain away the results of the M-M experiment. He said that the contraction of the arms of the interferometer (the machine they used) in the direction of travel made the interferometer arms shorter by just the amount needed to compensate for what was expected to be a longer travel time for the light through the moving ether. Even as late as 1931, Einstein still claimed that the ether existed. The theory of relativity was invented primarily to keep the idea of the ether alive by explaining why it was not detected in the M-M experiment.

In 1911, Max Planck published his famous "second paper" in which the existence of an all-pervasive vacuum energy (the Zero Point Energy) was proposed as the underlying reason for quantum uncertainty. However, this proposal was ignored for a number of years because of the focus on Planck's "first paper" published in 1901, which presented the idea of ZPE as a purely theoretical concept. This first paper gave rise to a new branch of physics: Quantum Electrodynamics (QED).

Then, in 1962, Louis de Broglie *("de Broyh")*, who had written one of those early papers that gave rise to QED, suggested that perhaps physics had missed something important, after all. He wanted his colleagues to re-examine Planck's second paper which stated that the Zero Point Energy was a real physical entity that caused the observed quantum effects. The re-examination of the reality of the ZPE did begin, and is still on-going. Physicists who support the idea that the ZPE is real, not theoretical, are called SED physicists.

The most famous equation associated with Einstein, E=mc², does not actually have anything to do with special relativity. The equation simply states the relationship between matter and energy, and is also used by SED physicists. Oddly enough, this equation probably did not originate with Einstein.

This expression appeared in the science magazine *Atti* in 1903, in an article by Olinto De Pretto. Einstein was undoubtedly aware of this fact. Expert mathematicians have analyzed Einstein's derivations for this formula and have found a number of inconsistencies and problems, causing them to suggest that he "forced" his derivations in order to make them match a formula he had previous knowledge of. Another science historian (Stark, 1907) said that Planck gave the first derivation for E-mc². However, Einstein's supporters carried the day with their adamant claims that Einstein's work was completely original.

It is generally believed that the theory of relativity has been proved (with no room for doubt) by experimental confirmation of predictions that the theory has made. However, all the major predictions of relativity also follow naturally from classical physics using the ZPE approach. Here are three of the best-known relativity "proofs" and their ZPE explanations.

1) Gravitational lensing

A gravitational lens occurs when there is a large concentration of matter between a galaxy, or star, and the earth. The massive object appears to bend the light coming from the galaxy or star, in the same way that a lens would. The relativity explanation is that massive objects have a lot of gravity, and gravity bends light. The greater the mass, the more it bends the "space-time continuum" in which everything exists. You've probably seen illustrations of the "gravity wells" around astronomical bodies, looking like a bowling ball sinking into a trampoline. It has been claimed that astronomers have observed examples of gravitational lensing, thus proving relativity to be true.

In fact, there are far too few gravitationally lensed objects compared to what relativity predicts. Among the few examples found, there are contrary examples, where the object bending the light does not seem to have enough mass to do so. In these cases, dark matter is invoked (as usual) as the answer.

Plasma astronomers agree that a strong gravitational field is needed; however, that is not the whole story (as indicated by the invention of dark matter by regular astronomers). Plasma also bends light. We can observe this to be true by looking at the plasma in the vicinity of our sun. Starlight that goes through this layer is strongly bent, as noted by E. H. Dowdye in his paper on this topic in the *Proceedings of the NPA*, No. 9, 2013. If we look for plasma filaments in and around the galaxies that relativity advocates use as their "proof," we find that often we are looking end-on at the filament in which the galaxy clusters are immersed. The lensing effects are seen most strongly around the outer edge of the filament since that is where a "double layer" of plasma occurs. Thus, the light comes to us after having traveled down the length of a plasma cylinder, or through its pinched region which can give arc-like distortions. Also, in a place where lensing should definitely occur—near the center of the Milky Way, where there is a supposed black hole—none has been found. The "black hole" is just a fast-spinning central plasmoid (that used to be a quasar) which still emits x-rays and other radiation.

2) The perturbed orbit of the planet Mercury (the advance of its perihelion)

Mercury traces an elliptical path around the sun. The closest point of that ellipse to the sun is called the perihelion. However, Mercury does not trace exactly the same path each time. Mercury's erratic orbit was a mystery until relativity applied three mathematical corrections, one for time dilation, one for space contraction, and one for mass increase due to speed. When these three effects were applied to the formula for the motion of the orbit perihelion, Mercury's eccentricity was accounted for.

However, the ZPE can also account for Mercury's behavior. In 1999, SED physicists suggested that the ZPE will be thicker closer to large bodies such as stars. Thus, planets that get close enough to the sun should show some effect of this greater density of ZPE, which will perturb their orbits.

3) The slowing of atomic clocks in gravitational fields

"GPS systems rely on relativity equations for their accuracy." Yes, these equations work. However, the underlying reason for why they work doesn't have to be time dilation. The strength of a gravitational field could be due to secondary EM radiation emitted by oscillating charges that make up all matter. More matter means more EM radiation, and more radiation means a greater ZPE density, and greater ZPE will slow down atomic processes like those used in atomic clocks.

For a more thorough discussion of these three points (often with many equations), and for additional points, please see chapter 7 of Setterfield's book, *Cosmology and the Zero Point Energy*.

BIBLIOGRAPHY

Barry Setterfield's textbook:
Cosmology and the Zero Point Energy. Natural Philosphy Alliance Monograph Series, No. 1, 2013.
  Published by the Natural Philosophy Alliance, Inc. ISBN 978-1-304-19508-1
  Available for purchase at: https://www.barrysetterfield.org/GSRdvds.html

Barry Setterfield videos I accessed via YouTube:
"Plasma Astronomy and the Bible" https://www.youtube.com/watch?v=-J6rbzuhsKM&t=2332s
"Lightspeed: A Journey of Discovery" https://www.youtube.com/watch?v=3umDkXPqwTA&t=2680s
"Sun Puzzles" https://www.youtube.com/watch?v=rezxQW06BhY
"Origin of the Elements:" https://www.youtube.com/watch?v=DkH0YprMKnY&t=1184s
"Genesis Science Research New Developments, videos parts 1-4"

Here are some examples of the primary sources from *Cosmology and the Zero Point Energy*. There are many more listed at the end of each chapter.

Alfven, H. *Cosmical Electrodynamics.* Oxford University Press, New York, 1950.
Arp, H. *Seeing Red: Redshifts, Cosmology, and Academic Science.* Aperion, Montreal, 1998.
Ashmore, L. "Hydrogen cloud separation as direct evidence of the dynamics of the universe" in:
    Potter, F. (Ed.), 2nd Crisis in Cosmology Conference, Port Angeles, WA, 2008, Astr. Soc. Pacific
    Conference Series, vol. 413, 3–11, 2009, vixra.org/pdf/1008.0074v1.pdf
Bearden, J., and H. M. Watts. "A Re-evaluation of the Fundamental Atomic Constants." Phys. Review, 81 (1951)
Boyer, T. H. *Phys. Rev.* D 11 (1975), p. 790.
De Broglie, L. *New Perspectives in Physics.* Basic Books Publishing C., New York, 1962.
Narlikar, J. and H. Arp. "Flat Sapcetime Cosmology: A Unified Framework for Extragalactic Redshifts."
    Astrophysical Journal 405:1, pp. 51-56.
De la Pena, L. "Stochastic Electrodynamics: Its Development, Present Situation, and Perspectives."
    Stochastic Processes Appliled to Physics and other Related Fields. World Scientific Publishing Co.,
    Pty. Ltd. pp. 428-451. 1983.
Gheury de Bray, M.E.J. *Isis*, vol. 25 (1936), pp. 437-448.
Haisch, B., A. Rueda, and H. E. Putoff. *Phys. Rev.* A 49, 1994, pp. 678-694.
Lerner, E.J. "Tolman test from z = 0.1 to z = 5.5: preliminary results challenge the expanding universe
    model" in: Potter, F. (Ed.), 2nd Crisis in Cosmology Conference, Port Angeles, WA, 2008; Astr. Soc.
    Pacific Conf. Series, vol. 413, 3–11, 2009, arxiv.org/PS_cache/arxiv/pdf/0906/0906.4284v1.pdf
Lerner, E.J., "Evidence for a non-expanding universe: surface brightness data from HUDF" in: Lerner, E.J.
    and Almeida, J.B. (Eds.), First Crisis in Cosmology Conference, AIP Conf. Proc. 822, AIP, p. 60, 2006
Peratt, A. L. *IEEE Transactions On Plasma Science*, PS-14 (6), 1986, pp. 763-778.
Puthoff, H. E. *Phys Rev.* A 40 (9): 4857 (1989)
Schilling, G. "Do gamma ray bursts always line up with galaxies?" Science 313:749, 2006
Spika, V. et al. *Beyond the Quantum*, edited by Theo M. Nieuwenhuizen (World Scientific, 2007), pp. 247-270.
Tejos, N. et al. "Casting light on the 'anomalous' statistics of MgII absorbers towards gamma-ray burst
    afterglows: the incidence of weak systems," *Astrophys. J.* 706:1309–1315, 2009
Tifft, W. G. *Astrophysical Journal*, 206:38 (1976), 211:31 (1977) and 211:377 (1977)
Verschuur, G. *Astrophysics & Space Science*, No. 227, pp. 187-198, 1985.
Wesson, P. S. *Cosmology and Geophysics; Monographs on Astronomical Subjects: 3*, pp.65-66, 88-89,
    115-122, 207-208. Adam Hilger, Ltd., Bristol, 1978.

https://plasmauniverse.info/
http://www.haltonarp.com/articles/intrinsic_redshifts_in_quasars_and_galaxies.pdf
http://www.astr.ua.edu/keel/galaxies/cmbr.html

Here are some additional resources I used:

How plasma balls work:
https://www.youtube.com/watch?v=w7dvJj5gB4g (High School Physics Explained channel)

What is plasma?
https://www.psfc.mit.edu/vision/what_is_plasma

Story of the Horn antenna
https://www.youtube.com/watch?v=lu2G01ehx68 (Nokia Bell Labs)
https://www.youtube.com/watch?v=-k5bR0JH_5k   (Naked Science)

CMBR
https://en.wikipedia.org/wiki/Cosmic_microwave_background
https://www.youtube.com/watch?v=PPpUxoeooZk   (PBS Space Time)
https://www.youtube.com/watch?v=3tCMd1ytvWg&t=324s (PBS Space Time)
https://www.youtube.com/watch?v=AYFDN2DSVgc (Fermi Lab)
https://www.balzan.org/en/prizewinners/paolo-de-bernardis-and-andrew-lange/
http://spaceref.com/news/viewpr.html?pid=4675

BOOMERanG
https://en.wikipedia.org/wiki/BOOMERanG_experiment

Star formation
https://www.cfa.harvard.edu/research/topic/star-formation
https://science.nasa.gov/astrophysics/focus-areas/how-do-stars-form-and-evolve

Meter-size barrier
https://astrobites.org/2015/04/03/what-is-the-meter-size-barrier/

Problems with star formation theory
https://cosmosmagazine.com/space/star-forming-theory-thrown-into-question/
https://www.realclearscience.com/articles/2013/09/04/big_problem_in_star_formation_theory_solved_106655.html
https://answersresearchjournal.org/astronomy/review-stellar-formation-theory/
https://en.wikipedia.org/wiki/James_Jeans

Distant galaxies
https://www.universetoday.com/156359/once-again-galaxies-look-surprisingly-mature-shortly-after-the-beginning-of-the-universe/
https://www.space.com/old-galaxy-in-early-universe-aless0731
https://en.wikipedia.org/wiki/GN-z11
https://en.wikipedia.org/wiki/Stellar_parallax
https://en.wikipedia.org/wiki/Coma_Cluster

Nebulas
https://en.wikipedia.org/wiki/Emission_nebula
https://en.wikipedia.org/wiki/Reflection_nebula
https://en.wikipedia.org/wiki/Planetary_nebula

Alignment of quasars
https://www.eso.org/public/news/eso1438/

Expansion
https://en.wikipedia.org/wiki/Expansion_of_the_universe
https://en.wikipedia.org/wiki/Hubble%27s_law
https://starchild.gsfc.nasa.gov/docs/StarChild/questions/redshift.html

Interpretation of redshifted galaxies
https://creation.com/our-galaxy-is-the-centre-of-the-universe-quantized-redshifts-show
http://www.cs.unc.edu/~plaisted/ce/redshift.html
https://www.ospublishers.com/The-Redshift-Blunder-has-been-Obstructing-Cosmology-for-Over-a-Century.html
https://beyondmainstream.org/other-explanations-for-red-shift/

Parallax
https://www.youtube.com/watch?v=2vPB8VmBdWU&t=756s (PhysicsHigh)

Cepheids
https://www.youtube.com/watch?v=BWs-ONRDDG4 (Khan Academy)
https://en.wikipedia.org/wiki/Cepheid_variable
https://www.youtube.com/watch?v=QcChCeX2VrY&t=180s (Science Online)

Supernovae
https://www.youtube.com/watch?v=ljoeOLuX6Z4 (Veritasium)
https://www.youtube.com/watch?v=BHkkSxIZ0v4 (PhysicistMichael)
https://www.youtube.com/watch?v=cJLMB5OUpqQ (jason Kendall)

Blackbodies and Planck's constant
https://www.youtube.com/watch?v=97AqZBjcdUM (The Action Lab)
https://www.youtube.com/watch?v=7hxYGaegxAM (PhysicsHigh)
https://www.youtube.com/watch?v=tQSbms5MDvY (PBS Space Time: Planck's Constant and QM)
https://www.youtube.com/watch?v=jw1w5ijQyvg&t=309s (Catalyst University)
https://www.youtube.com/watch?v=X5rAGfjPSWE (PBS Space Time: Nature of Nothing)

Planck and Planck's constant
https://www.britannica.com/science/Plancks-constant
https://www.techtarget.com/whatis/definition/Plancks-constant
https://science.howstuffworks.com/dictionary/physics-terms/plancks-constant.htm

Discussion of evidence for amd against expanding universe
https://creation.com/expanding-universe-2
http://www.cs.unc.edu/~plaisted/ce/redshift.html
https://faculty.humanities.uci.edu/bjbecker/ExploringtheCosmos/lecture20.html

Research challenging the cosmological expanding universe theory (standard model)
https://www.universetoday.com/146523/evidence-is-building-that-the-standard-model-of-the-expansion-of-the-universe-needs-some-new-ideas/

Casimir effect
https://www.youtube.com/watch?v=OgJj49ws478 (See the Pattern channel)
https://en.wikipedia.org/wiki/Van_der_Waals_force
https://en.wikipedia.org/wiki/Casimir_effect

Electron wave patterns
https://www.youtube.com/watch?v=4qxfBTqBFn8
https://www.youtube.com/watch?v=7V6j4sZ7w_E

Trying to detect ZPE experimentally
https://www.scientificamerican.com/article/something-from-nothing-vacuum-can-yield-flashes-of-light/
https://scitechdaily.com/clever-physics-experiment-that-produces-something-from-nothing/

Black holes
https://astronomy.com/magazine/2019/08/a-brief-history-of-black-holes
https://www.youtube.com/watch?v=-FdWTH08u30 (Thunderbolts video)

Articles (not by plasma scientists) reporting electric currents, (filaments) in outer space:
https://www.mdpi.com/2075-4434/5/4/71 (Research by Ioannis Contopoulos)
https://www.quantamagazine.org/the-hidden-magnetic-universe-begins-to-come-into-view-20200702/
https://iopscience.iop.org/article/10.3847/2041-8213/aa9985/meta (Astrophysical Journal Letters, 2017)
https://news.northwestern.edu/stories/2022/01/nearly-1000-mysterious-strands-revealed-in-milky-ways-center/

I also consulted one of the most popular secular books on plasma astronomy:

Scott, Donald E. *The Electric Sky*. Milamar Publishing, Oregon, 2012. ISBN 978-0-9830966-6-5

CPSIA information can be obtained
at www.ICGtesting.com
Printed in the USA
JSHW042141280623
43835JS00003B/7